THE BONES & MUSCLES

YOUR BODY YOUR HEALTH

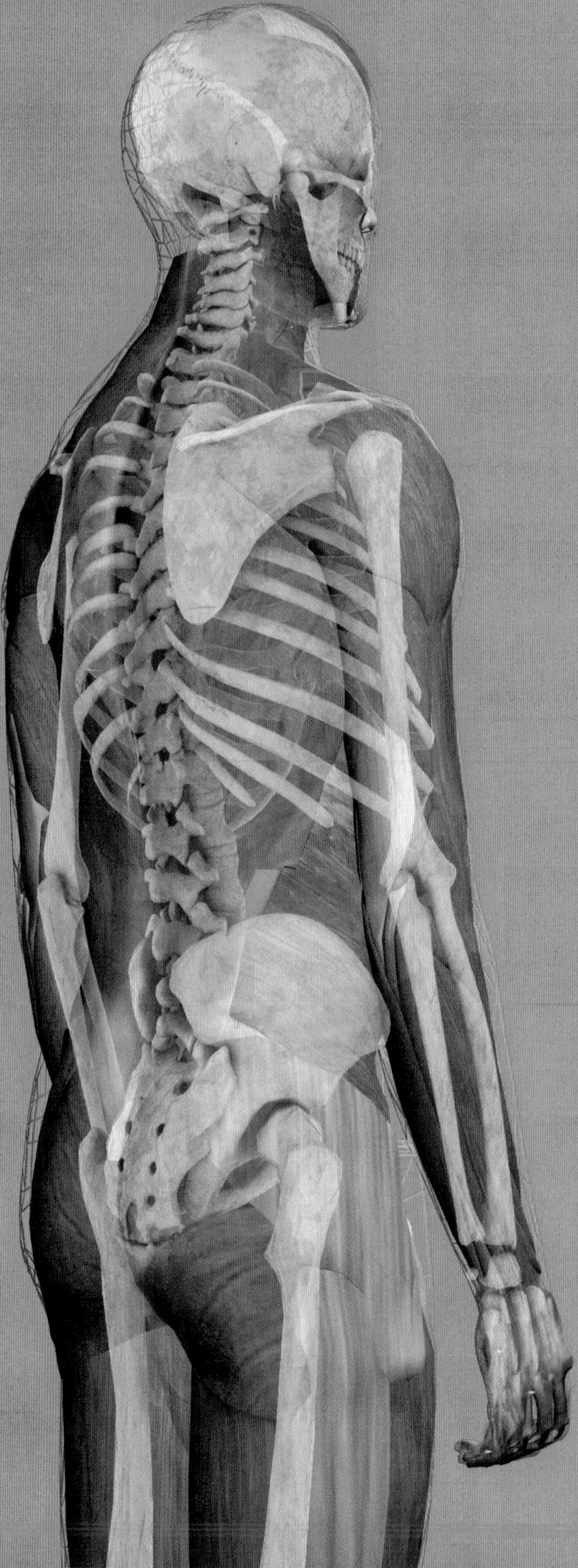

YOUR BODY YOUR HEALTH

THE BONES & MUSCLES

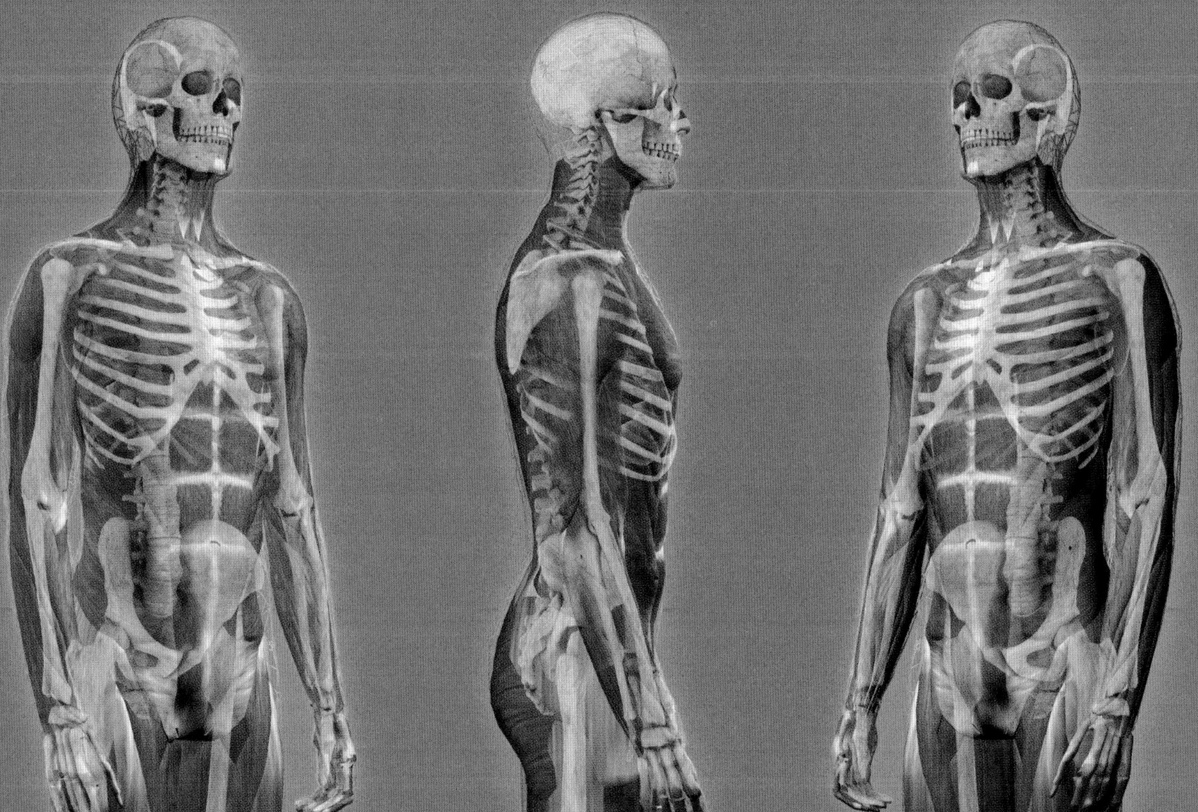

The Reader's Digest Association, Inc.
Pleasantville, New York
London New York Sydney Montreal

The Bones and Muscles

was created and produced by
Carroll & Brown Limited
20 Lonsdale Road
London NW6 6RD
for Reader's Digest, London

Library of Congress Cataloging-in-Publication Data
The bones and muscles / Reader's Digest.--
1st American ed.
p. cm. -- (Your body, your health)
Includes index.
ISBN 0-7621-0485-6 (hardcover)
1. Musculoskeletal system--Popular works.
I. Reader's Digest Association. II. Series.
QP301.B55 2003
611'.71--dc22
2003016008

Printed in the United States of America
1 3 5 7 9 8 6 4 2

The information in this book is for reference only; it is not intended as a substitute for a doctor's diagnosis and care. The editors urge anyone with continuing medical problems or symptoms to consult a doctor.

American Edition Produced by
NOVA Graphic Services, Inc.
2370 York Road, Suite A9A
Jamison, PA 18929 USA
(215) 542-3900

President
David Davenport

Editorial Director
Robin C Bonner

Composition Manager
Steve Magnin

Associate Project Editor
Linnea Hermanson

Otolaryngology Specialist Consultant
Dr Milan Amin, MD
Drexel University College of Medicine/Hahneman Medical College, Philadelphia

CONTRIBUTORS

Wynnie Chan, BSc, PhD, Public Health Nutritionist

Dr Sandeep H Cliff, MB, BSc, MRCP, Consultant Dermatologist, Honorary Senior Lecturer, Surrey

Mr Brian Coghlan, MD, FRCS (Plast), Consultant Plastic Surgeon, Guy's Hospital, London

F J Cunningham, MRIPH MIT (Lond), Member of the Institute of Trichologists, The Trichology Centre, Rochdale

Christel Edwards-de Graauw, Nail Technician, Nail Artist and Makeup Artist, Fingernails Direct, Northern Ireland

Dr Colin Fleming, BSc, MB ChB, MRCP, Consultant Dermatologist, Department of Dermatology, Ninewells Hospital, Dundee, Honorary Senior Lecturer, University of Dundee

Katy Glynne, BSc, MRPharmS, Dip Pharmacy Practice, Clinical Services Manager, Charing Cross Hospital, London, Clinical Lecturer, The School of Pharmacy, University of London

Dr Lesley Hickin, MB BS, BSc, DRCOG, MRCGP, General Practitioner

Dr Shona Ogilvie, MB ChB, MRCGP, Clinical Fellow, University Department of Dermatology, Ninewells Hospital, Dundee

Penny Preston, MB ChB, MRGCP, Medical Writer

Beverly Westwood, RN, BSc, MSc, Research Director to Mr Brian Coghlan

For Reader's Digest
Editor in Chief and Publishing Director Neil E Wertheimer
Managing Editor Suzanne G Beason
Production Technology Manager Douglas A Croll
Manufacturing Manager John L Cassidy
Production Coordinator Leslie Ann Caraballo

YOUR BODY YOUR HEALTH

The Bones and Muscles

Awareness of health issues and expectations of medicine are greater today than ever before. A long and healthy life has come to be looked on as not so much a matter of luck but as almost a right. However, as our knowledge of health and the causes of disease has grown, it has become increasingly clear that health is something that we can all influence, for better or worse, through choices we make in our lives. *Your Body Your Health* is designed to help you make the right choices to make the most of your health potential. Each volume in the series focuses on a different physiological system of the body, explaining what it does and how it works. There is a wealth of advice and health tips on diet, exercise and lifestyle factors, as well as the health checks you can expect throughout life. You will find out what can go wrong and what can be done about it, and learn from people's real-life experiences of diagnosis and treatment. Finally, there is a detailed A to Z index of the major conditions which can affect the system. The series builds into a complete user's manual for the care and maintenance of the entire body.

Your skeleton and muscles literally keep you upright and mobile. In this volume you will read about the structure of these vital body elements and how they function. You will see how and where bone is made, and how it interacts with muscles, tendons, ligaments and cartilage to facilitate your body's tremendous range of movement. You will discover the best ways to look after bones and muscles – from the unborn child right through to old age – to avoid both the common or more severe problems that can be so detrimental to quality of life. Find out which foods are good, and why weight control and exercise are so crucial. Learn about the risk factors and the simple steps you can take to help avoid them. Finally, discover what can be done to diagnose and treat any problems that do arise; the therapies available to counter pain, injuries and the more serious effects of aging; and the ingenious technologies employed by doctors today to correct physical deformities and replace damaged parts.

Contents

3

What happens when things go wrong

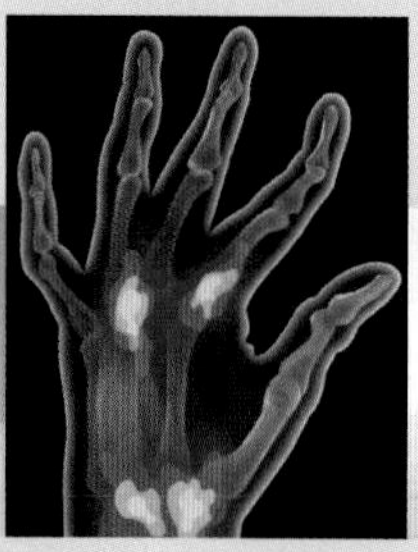

The life story of the bones and muscles

The story of evolution is written in bone—we trace its course down through the ages in the bones that our ancestors and other creatures have left behind. But the story of bone would be meaningless without muscle: Together, bones and muscles form a partnership that has enabled animals to conquer the sea, land, and air, and has made us what we are today.

It is all too easy to take for granted the tissues that give form and force to your body: your bones and the muscles that clothe them. But their remarkable properties and abilities, which make them the envy of every engineer and scientist, lie at the very heart of our humanity. We depend on them to give us shape and physical power, to allow us to react to and interact with the world around us, and to guide and work most of the systems of our bodies. It's also thanks to bones and muscles that we first became human.

Evolutionary throwback?
Some of the first bones took the form of hard scales embedded in the skin, and inside the animals' bodies, cartilaginous skeletons gave shape and structure. Today's sharks preserve some of these features, with skins covered in tiny teeth and soft, gristly skeletons.

DEM BONES

The first animals on Earth used primitive silicon-based structures to help stiffen their bodies. In the Cambrian period, about 500 million years ago, the ancestors of the vertebrates—creatures that had the ability to extract calcium from their environment and use it in the manufacture of bones—appeared. Animals before this had had muscles, but only now did they possess the knockout combination that was to make them rulers of the Earth: muscles in conjunction with a bony skeleton.

With a hard skeleton for their muscles to press against, the first vertebrates were able to make full use of their muscle power, becoming fast, agile swimmers, and bony plates on their heads developed into fearsome jaws and teeth. With these adaptations, they conquered the oceans. As animals developed still harder internal skeletons, they were able to explore new territory: the land. Away from the supporting embrace of the ocean, the pull of gravity made a stiff skeleton and strong muscles even more important. Today, your skeleton and muscles still play this crucial role—resisting gravity, supporting your organs, and allowing you to move freely.

By age 70, most of us will have lost 30 percent of our strength because of reduction in muscle mass. Some of this is reversible: An 8-week strength-training program for frail 80-year-olds resulted in a 174 percent increase in muscle mass and a 48 percent increase in walking speed.

TWO LEGS BETTER

Millions of years after the first land animals emerged from the sea, an apelike creature took the first steps on a road that led away from the other apes and toward us. That creature was the first hominid, and the crucial change that set it on the path to humanity concerned its bones and muscles. The change was the adoption of bipedalism: walking on two legs. Scientists are divided over the reasons that our ancestors began walking on two legs, but all agree that the consequences were profound. In order to support this new and unusual style of getting around, the human skeleton and muscles had to change so that the head and upper body were balanced directly on top of the legs. The spine shortened and the vertebrae became larger. The long bones of the thigh angled in toward the knees. The knee joint evolved the ability to lock in place. Large muscles developed on the sides of the hips to hold the body stable while walking. Most important of all, the pelvis narrowed to help align the legs directly underneath the upper body.

A narrower pelvis meant that babies had to be smaller at birth in order to fit through the birth canal. Because they started out smaller, the babies had to spend longer growing up, and this in turn had profound consequences for humanity. With longer childhoods came family and social units and the ability to pass acquired information from one generation to the next—in other words—learning. Several essential human characteristics have thus come about because of changes to our bones and muscles.

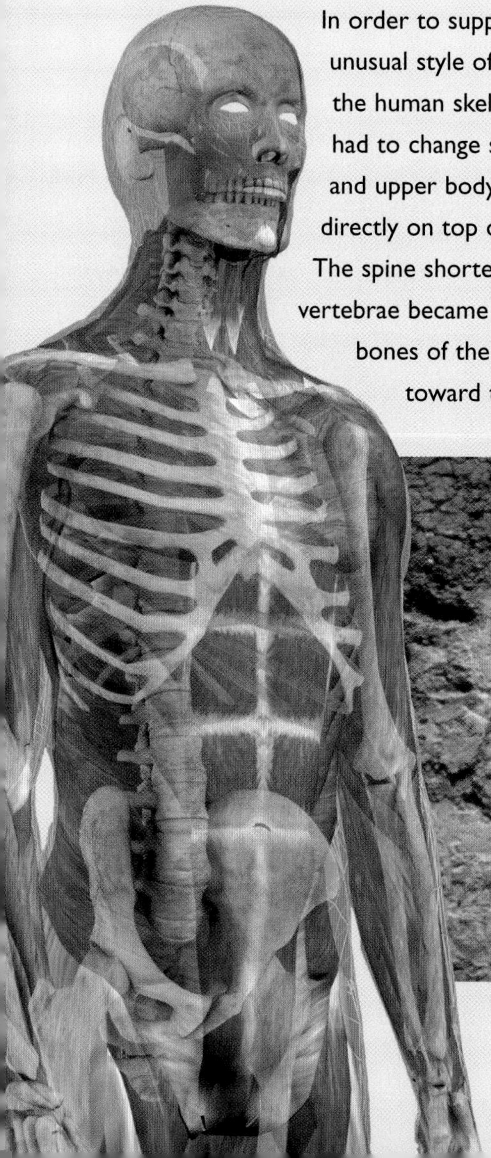

Walk this way

These footprints, at Laetoli in Tanzania, were made in newly fallen volcanic ash about 3.5 million years ago. They are the earliest known record of hominids walking upright, which is indicated by the deep impressions formed by the heel and big toe, the longitudinal arch, and the big toe being parallel to the other digits.

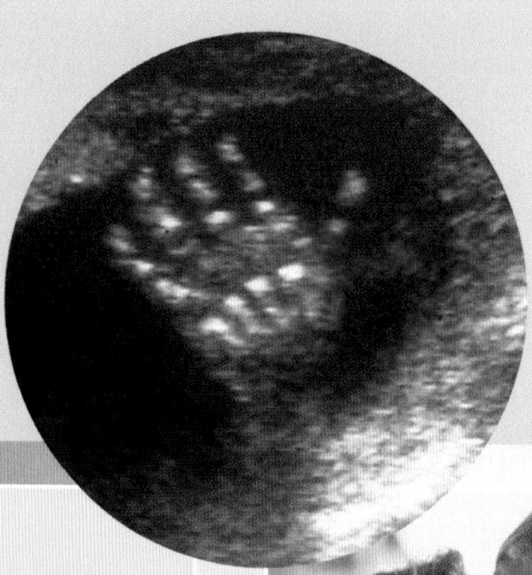

Hand bone
By 21 weeks, the age of the fetus in this ultrasound, the fingers are separated and bone is developing in the fingers and palm (white areas).

MULTIFUNCTIONAL TISSUE

Bones and muscles have many vital functions other than giving you the ability to walk around on two feet. They give you the power to move and manipulate objects, they shift fluids around inside your body (the heart, for instance, is one big muscle for pumping blood), they store calcium and other minerals, they protect your soft organs and blood-producing bone marrow, they generate body heat, and they allow you to communicate and express yourself.

THE LIFE OF THE BONES AND MUSCLES

Muscles and bones are generally resilient and strong. If well built early in life and strengthened and exercised in later years, they should allow a whole range of movement well into old age, without pain or discomfort.

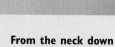

BABIES

HOW THE BONES AND MUSCLES DEVELOP

Your musculoskeletal system wasn't always this versatile or functional. When you were born, it was weak and vulnerable, even though it had started to develop almost from conception. A newly formed embryo starts off as a ball of cells, but within two weeks these cells differentiate into different types of tissue, including somites—blocks of tissue that develop into muscle. While this is happening, cartilage is being laid down to form "models" for the bones—like templates that attract bone-forming cells.

By the time the embryo is four weeks old, the vertebrae start to appear and the heart muscle starts to work, beating out the rhythm that it maintains for decades to come. By week 9, the entire skeleton is laid out in template form and the process of ossification begins. This is where calcium is laid down to produce the hard material we normally think of as bone. When the fetus is between 13 and 16 weeks old, it can make a fist and the leg muscles have developed enough to start kicking. After this, the baby grows steadily, building up bone and muscle mass until it is time to emerge into the world.

From the neck down
Babies gain control of their bodies from the top down, so the neck muscles support the weight of the head, and then the back and stomach muscles allow the baby to sit. Finally, the legs support the baby's body weight.

RUBBER LEGS

The musculoskeletal system still has a lot of developing to do after birth. Although a newborn's grip is strong enough to support the weight of the whole body, the neck muscles are not strong enough to lift the head and the leg muscles aren't strong enough to stand up. Even if they were, the leg bones would be too rubbery to support the weight.

One reason they remain so soft is to make it easier to squeeze through the birth canal. The bones of the skull do not fuse together until several weeks after birth, so that during labor they can slide over one another, allowing the baby's skull to pass through the mother's pelvis. Not until a baby is about a year old are the legs strong enough to support the body.

Hop, skip, and jump
Children like and need exercise to build strong bones and muscles, but the number and nature of their activities make them a high-risk group for bone fractures. Outdoor activities ensure they get enough vitamin D, which is essential for calcium absorption.

Milk marvelous milk
Studies have shown that increasing calcium intake can improve bone density. All dairy products are good sources, as are leafy green vegetables. Skim milk is lower in calories and higher in calcium than whole milk; some soy milks are fortified with the mineral.

CHILDREN

ADULTS

IRON MAN (AND WOMAN)

During childhood, bones continue to get thicker and harder, but they remain soft at the ends so that they can grow longer. Muscle mass slowly increases, too. Most bone and muscle growth occurs during puberty and the early teens, and by age 17 about 90 percent of final bone mass is present. By now the bones are as strong as iron but three times lighter, and the jaw muscles used for chewing are strong enough to support the entire body weight.

CONSTANT REMODELING

Not until the early 20s do the soft ends of the bones finally harden, permanently fixing the length of the bone. But bones can still get thicker or thinner or even change their shape in response to how they are used. This is because of a process called bone remodeling, in which mineral content of the bone—mainly calcium—is constantly being recycled and replaced. Between 10 and 30 percent of the adult skeleton is replenished each year. This helps maintain the optimal levels of calcium in the blood and the structural function of the skeleton.

Although remodeling continues throughout life so that bone shape can always change, overall bone mass generally reaches its peak in the 20s. After this, bone is very gradually lost, a process that accelerates in old age, and, for women, particularly after menopause, when estrogen production stops.

Muscle mass also decreases with old age, but research on elderly nursing home residents shows that load-bearing exercise can increase muscle mass and improve strength even in this age group, proving that it's never too late to start exercising!

BODY WORKS

Bone and muscle loss may not happen until later in life, but you can take action to prevent it from the very beginning of life. Although the size and mass of your bones and muscles are partly predetermined genetically, other

Don't stop moving
Independence and quality of life are directly linked to muscle strength. Both bones and muscles benefit from weight-bearing exercise at any age.

OLDER PEOPLE

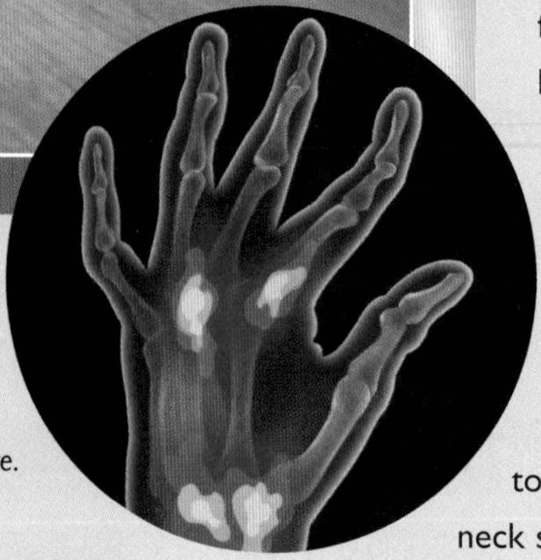

The color of pain
In this X-ray of an elderly hand, arthritic joints of the fingers and wrist are false-colored yellow. Red areas signify pain caused by swelling and joint damage.

factors played a crucial part in shaping them and continue to do so right now. Your mineral and vitamin levels, particularly of calcium and vitamin D, have a major impact on bone mass, and muscle mass and composition is largely determined by your activity levels.

BONE APPETITE

The earlier you boost your calcium intake and start getting exercise, the healthier your bones and muscles will become, and the healthier they'll stay: Research suggests that for every 5 percent increase in bone density during childhood and adolescence, the risk of later fracture falls by 40 percent. Lifestyle factors are also important. Smoking, poor eating patterns, and excess alcohol use can all detract from bone mass, and too much salt in the diet may increase the amount of calcium lost in the urine, so you need to boost the amount of calcium you eat to keep levels stable.

BREAKING POINT

Daily life puts a lot of stress on your musculoskeletal system, and many leisure pursuits, such as sports, test bones and muscles to the limit. Although they have evolved remarkable strength and endurance for their size and weight, even the healthiest bones and muscles inevitably suffer to some extent under this constant strain. Surveys show that more than 80 percent of Americans have jobs that involve sitting for extended periods—a posture that the spine did not evolve to cope with. As a result, back and neck strain are among the most common ailments affecting people in the developed world. Meanwhile, for people under age 30, bone fractures and muscle strains are the most common serious ailments, period. Furthermore, in the developed world, a sedentary lifestyle is likely to contribute to an individual being overweight, placing additional strain on the bones and joints.

FROM X-RAYS TO HRT

Medical science has developed an array of technologies to diagnose and treat bone and muscle problems. The development of X-rays, first discovered in 1895, ranks alongside the invention of vaccination and the discovery of penicillin as one of the great achievements of early modern medicine. Since then, a host of sophisticated scanning technologies, such as magnetic resonance imaging, have enabled doctors to diagnose musculoskeletal problems with increasing accuracy. The setting of bones and relief of strains have been staples of the physician's trade for cenuries, but new approaches such as titanium pins, bone grafts, artificial joints, and complex microsurgery give doctors the power to repair many serious bone, muscle, and joint problems. In recent years, new drugs and the introduction of hormone replacement therapy have radically improved the prognosis for osteoporosis.

TURNING BACK THE CLOCK

One of the most exciting approaches currently in development is stem-cell technology. In this, chemicals are used to turn back the clock in a sample of a person's cells, so that they revert to an earlier stage of development. These cells can then be harnessed to grow new parts of the individual's body to replace old or damaged areas.

A promising application of this new technology is in the field of bone grafts and treatments for muscle, tendon, and ligament injuries. In the near future, it may be possible to repair even badly broken bones and torn muscles by extracting a cell sample from the patient, growing it in a laboratory, and then reinserting it in the site of the injury, where it will quickly heal the affected area.

ARTIFICIAL ARMS AND JUICED UP JOINTS

Researchers at the University of Washington are currently working on another approach to solving bone and muscle problems: They are building an artificial human arm, using lightweight metal bones and joints, and a form of artificial muscle that uses tiny amounts of compressed air to mimic the abilities and properties of the real thing. In the future, such technologies might lead to workable, responsive artificial bones and muscles for amputees and paraplegics, and might even enhance the body's normal abilities.

Similar ambitions may be realized when genetic engineering becomes a reality. By tweaking the genetic code, it may be possible not only to cure hereditary bone, joint and muscle disorders but also to prevent degenerative problems like osteoporosis, and even to enhance the strength and endurance of human muscle and bone. Most scientists would agree, however, that it will be a long time before we can hope to improve on the remarkable properties and abilities of the amazing musculoskeletal system.

Breakthrough bone

Research on bone implants is ongoing. Liquid glass (right) can be injected into a complex fracture or osteoporotic bone. Bone cells will adhere to the glass and grow and bind together to form new tissue. This procedure is uncommon in the United States

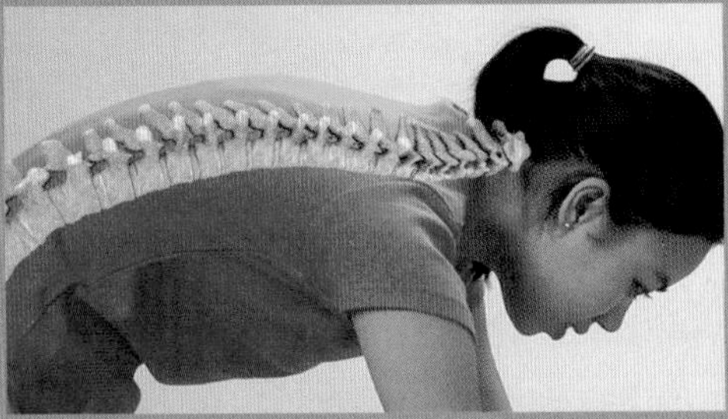
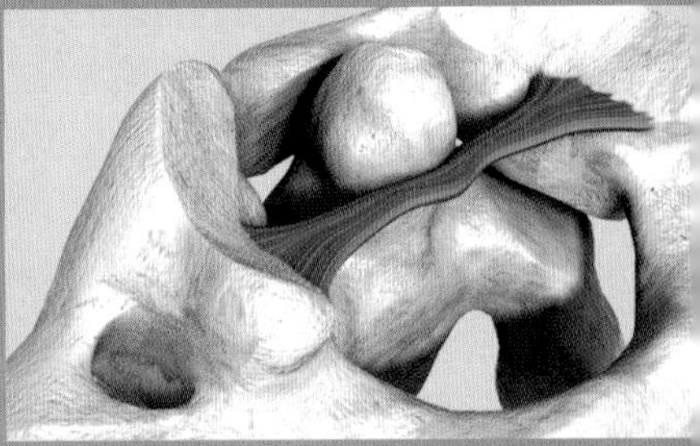

1

How your bones and muscles work

Your amazing bones and muscles

The skeleton forms the framework that gives the body its basic shape. Muscles work with the skeleton—a bit like a system of levers and pulleys—to enable the body to perform all of its remarkable variety of movements and to support its other tissues and organs.

A STRONG FRAMEWORK

The bones of the skeleton and the groups of muscles that move them are attached around the joints, the hinges that connect bones together. Not all bones of the skeleton move, however. Some, such as those that make up much of the skull, are fused to form rigid structures.

In the following pages, we will look at the structure of the skeleton and the bones that form it, some of which may surprise you. Then we will examine in more detail different parts of the skeleton and the types of bones, how they are made, and what happens when they are damaged. We will also look at the different types of joints, some of which allow bones to move freely and others that restrict movement or allow no movement at all.

MUSCLES FOR MOVEMENT

Bones may provide the body's internal scaffolding, but they would not even enable you to stand upright—let alone move—were it not for the support and action of the skeletal muscles that surround them. These muscles extend across joints to produce movement of the bones they connect. This movement is achieved by contraction, shortening and thickening of the muscles, which is activated by the nervous system and under our conscious control. We look in detail at the structure of the skeletal muscles and how they produce movement. We also look briefly at the two other types of muscle in the body: the cardiac muscle in the wall of the heart and the smooth muscle found in the walls of blood vessels, the digestive system, and some other internal organs.

GLOSSARY

Bone *The hard, dense connective tissue that forms the skeleton of the body.*

Muscle *Tissue with cells that have the ability to contract to produce movement or force.*

Cartilage *Gristlelike, smooth, slippery tissue that coats the moving surfaces of joints.*

Ligament *A tough, fibrous band of connective tissue that links two bones together at a joint. A ligament plays an important role in keeping a joint stable.*

Tendon *A fibrous cord of connective tissue that joins a muscle to a bone. Tendons are key to muscle and joint movement.*

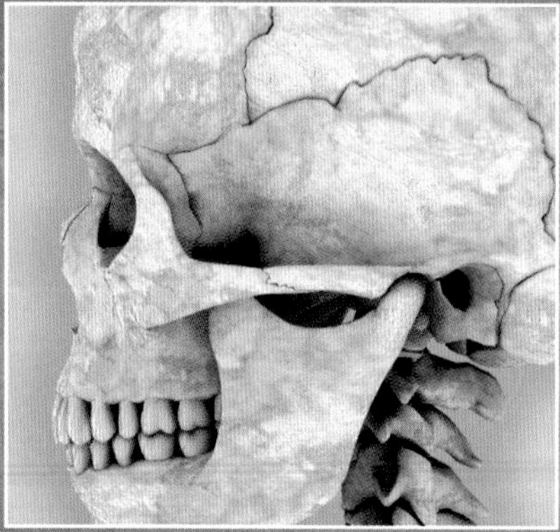

The skull *encases the brain and protects it from damage.* *For more information on the bones of the skull, see pages 20–21.*

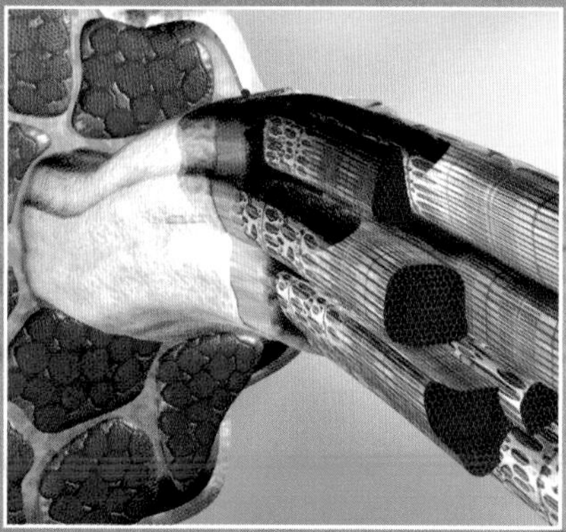

Skeletal muscles *are made up of bundles of long fibers that contract to produce movement.* *To discover more, turn to pages 32–35.*

The biceps muscle *extends across the elbow joint and enables the arm to bend. To read about other skeletal muscles in the body, see pages 30–31.*

The rib cage *protects the contents of the chest. Other functions of the skeleton can be found on pages 18–19.*

The humerus *in the upper arm is one of the long bones of the body. To read about its structure and how bones are made, see pages 24–27.*

The vertebral column *supports the body and protects the spinal cord. Its flexibility allows us to bend forward, backward, and from side to side. It is described in detail on pages 22–23.*

The femur *is the longest bone in the body; fractures are common. How they heal is discussed on pages 26–27.*

Muscles and bones *work together to make possible every move that you make. Find out more about how this happens—and the crucial roles of the brain and nervous system—on pages 34–35.*

The knee *is a hinge joint, allowing movement in two directions. To find out about this and other types of joints in the body, see pages 28–29.*

The skeleton accounts for about one fifth of the average person's body weight.

The skeleton

Within the intricate framework of the skeleton is an ingenious array of bones of various shapes and sizes. The resulting structure is strong enough to support the rest of the body yet is also able to bend and flex.

The skeleton *is divided into two main parts.*
- *The axial skeleton comprises the skull, backbone, and ribcage.*
- *The appendicular skeleton comprises the arms and legs and the bones that connect these to the axial skeleton, namely the shoulder girdle (shoulder blades and collarbones) and the pelvic girdle.*

DIVERSE ROLES

In addition to enabling the body to stand upright and to move, the skeleton helps protect the body's organs and other internal structures. A less obvious function is the storage of certain necessary elements, calcium and phosphorus, in particular. Some long bones are also the site for the formation of red blood cells and some white blood cells.

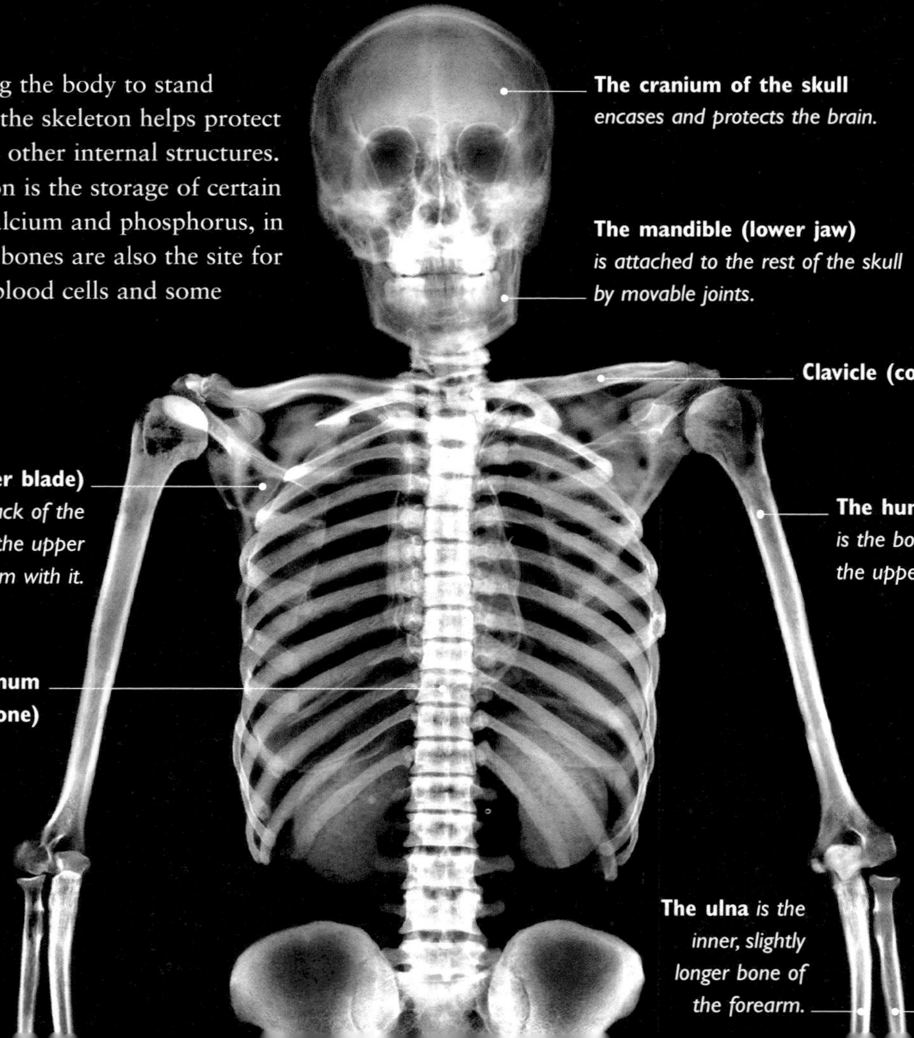
The cranium of the skull *encases and protects the brain.*
The mandible (lower jaw) *is attached to the rest of the skull by movable joints.*
Clavicle (collarbone)
The scapula (shoulder blade) *can slide over the back of the upper ribs, carrying the upper arm with it.*
The humerus *is the bone of the upper arm.*
Sternum (breastbone)
The ulna *is the inner, slightly longer bone of the forearm.*
The radius *is the outer, slightly shorter bone of the forearm.*

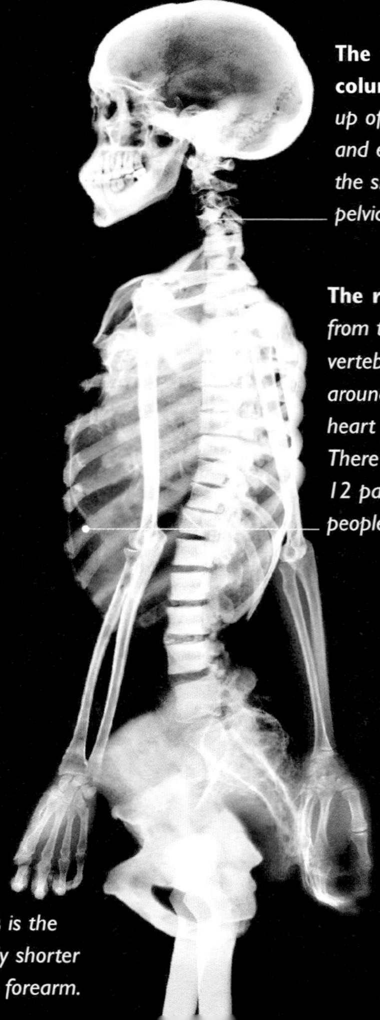
The vertebral column *is made up of 33 vertebrae and extends from the skull to the pelvic girdle.*
The ribs *project from the sides of the vertebrae and curve around to encase the heart and lungs. There are generally 12 pairs of ribs (a few people have 13 pairs).*

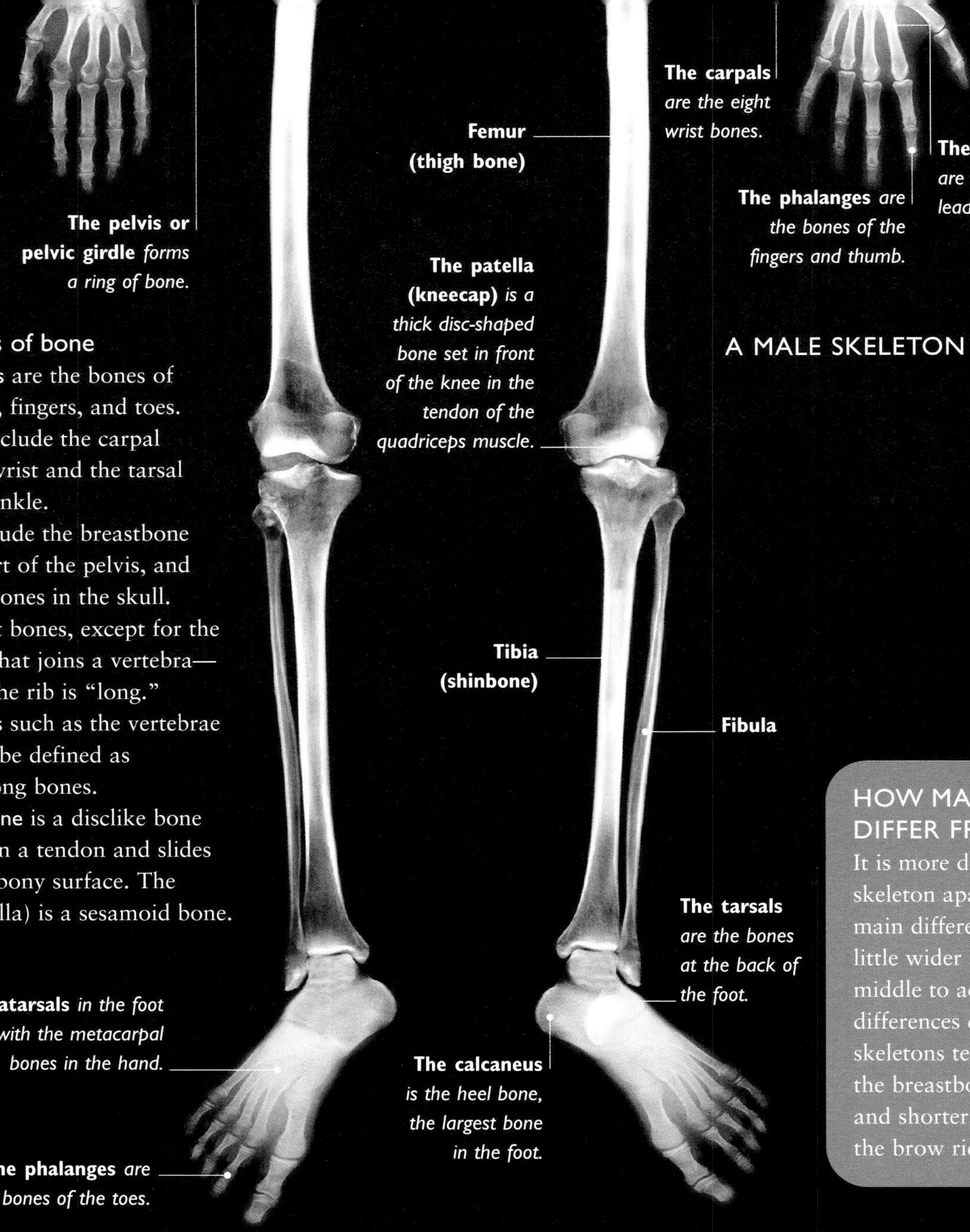
The pelvis or pelvic girdle forms a ring of bone.
Femur (thigh bone)
The patella (kneecap) is a thick disc-shaped bone set in front of the knee in the tendon of the quadriceps muscle.
Tibia (shinbone)
Fibula
The carpals are the eight wrist bones.
The phalanges are the bones of the fingers and thumb.
The metacarpals are the five bones that lead to the fingers.
The tarsals are the bones at the back of the foot.
The calcaneus is the heel bone, the largest bone in the foot.
The metatarsals in the foot correspond with the metacarpal bones in the hand.
The phalanges are the bones of the toes.

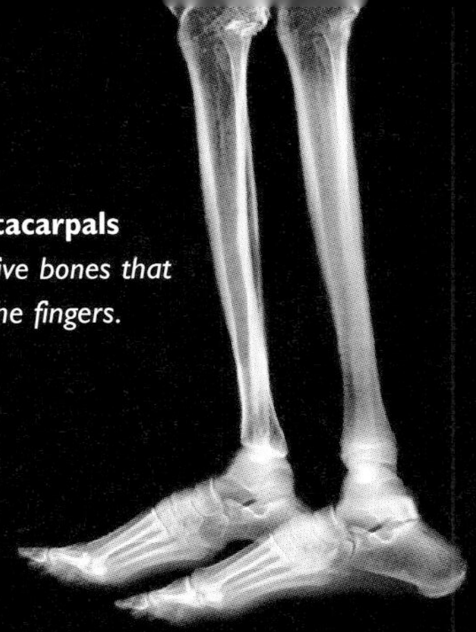

A MALE SKELETON

Different types of bone

- **The long bones** are the bones of the arms, legs, fingers, and toes.
- **Short bones** include the carpal bones in the wrist and the tarsal bones in the ankle.
- **Flat bones** include the breastbone (sternum), part of the pelvis, and many of the bones in the skull.

The ribs are flat bones, except for the part of the rib that joins a vertebra—this section of the rib is "long."

- **Irregular bones** such as the vertebrae can generally be defined as compressed long bones.
- **A sesamoid bone** is a disclike bone that lies within a tendon and slides over another bony surface. The kneecap (patella) is a sesamoid bone.

There are 206 bones in the human body. Of these, more than half are found in the hands and feet.

HOW MALE AND FEMALE SKELETONS DIFFER FROM EACH OTHER

It is more difficult to tell a male and female skeleton apart than you might imagine. The main difference is that the female pelvis is a little wider and has a larger opening in the middle to accommodate childbirth. Other differences do not always apply: Female skeletons tend to be smaller overall; in a female the breastbone (sternum) is commonly wider and shorter; and the jaw may be smaller and the brow ridges on the skull less heavy.

The skull

The delicate tissues of the brain require protection from the outside world. This is provided by the skull, a remarkable, mainly rigid structure that also plays a role in hearing and seeing and forms the structure for the face.

THE BONES OF THE SKULL

The skull is made up of two parts. The cranium encases the brain and provides resistance to external forces. The rest of the skull forms the face, housing the eyes, the ears, and the structures within the mouth. The bones have many tiny holes through which nerves, blood, and lymph vessels pass. There are 22 bones in the skull altogether, 8 of which are fused together to form the cranium.

Air-filled cavities

Within the skull are four pairs of air-filled cavities that are connected to the nasal cavity by passages. These are the frontal, maxillary, ethmoidal, and sphenoidal sinuses; each pair of sinuses is named after the part of the skull in which it lies.

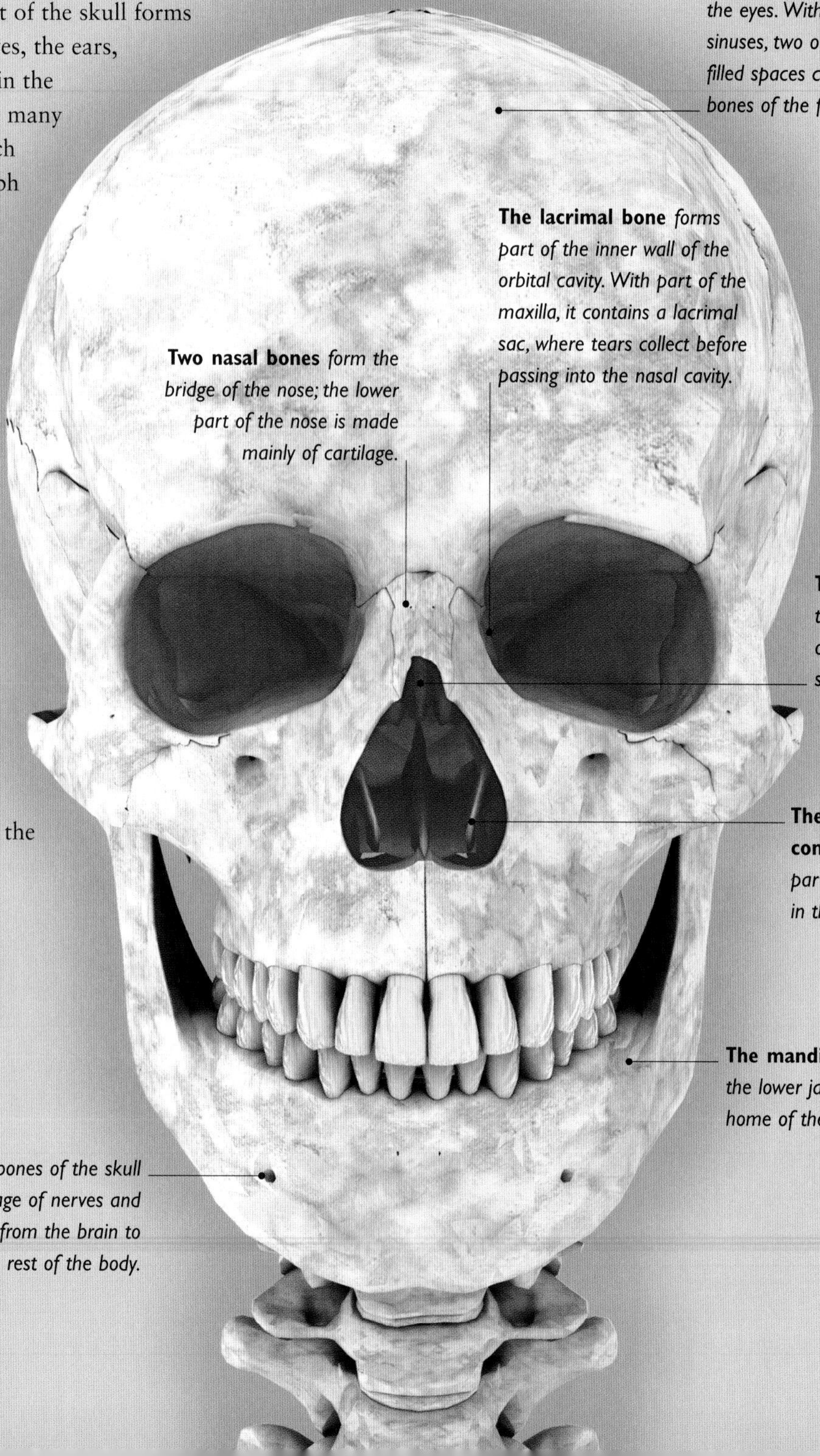
The frontal bone *forms the forehead and the upper part of the orbital cavities, which contain the eyes. Within it lie the frontal sinuses, two of a number of air-filled spaces contained within the bones of the face.*
The lacrimal bone *forms part of the inner wall of the orbital cavity. With part of the maxilla, it contains a lacrimal sac, where tears collect before passing into the nasal cavity.*
Two nasal bones *form the bridge of the nose; the lower part of the nose is made mainly of cartilage.*
The vomer *is the bony part of the nasal septum.*
The inferior nasal conchae *form parts of the bones in the nasal cavity.*
The mandible *forms the lower jaw and is the home of the lower teeth.*
Holes *in the bones of the skull allow for the passage of nerves and blood vessels from the brain to the rest of the body.*

SOFT SPOTS IN BABIES

At birth, the bones of the cranium have not yet fused, so there are spaces between the sutures that run across the skull that are known as "soft spots." These soft spots allow some flexibility of the skull during childbirth. The biggest of the soft spots, the anterior fontanelle, lies toward the front of the cranium. It fuses by the time a child is 1 year to 18 months old. The posterior fontanelle closes after about 3 months.

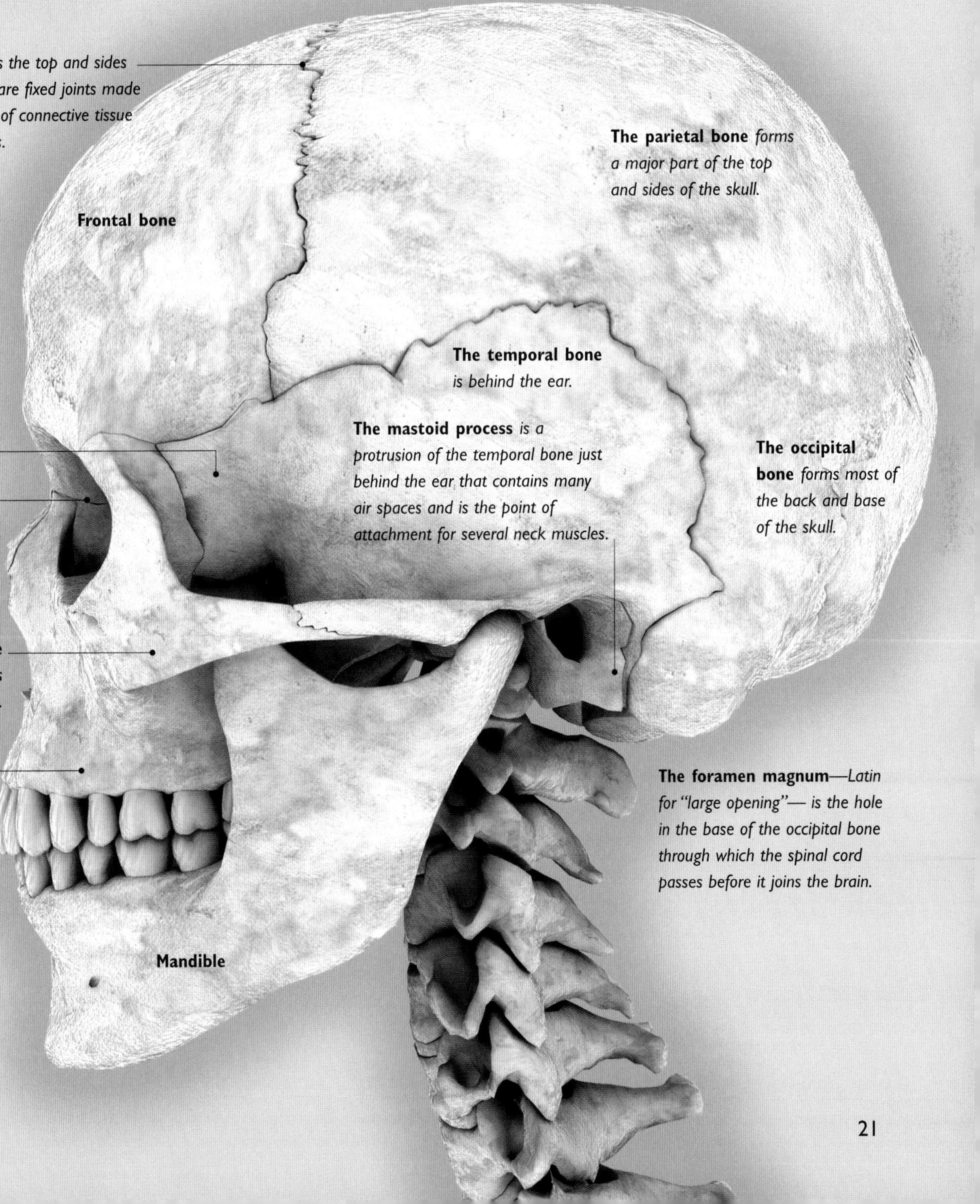

Sutures *run across the top and sides of the skull. These are fixed joints made of a small amount of connective tissue between two bones.*
The parietal bone *forms a major part of the top and sides of the skull.*
Frontal bone
The sphenoid bone *has a number of openings. One is the optic canal, which allows the optic nerve to pass from the eye toward the brain.*
The temporal bone *is behind the ear.*
The mastoid process *is a protrusion of the temporal bone just behind the ear that contains many air spaces and is the point of attachment for several neck muscles.*
The occipital bone *forms most of the back and base of the skull.*
The ethmoid bone *forms the inner wall of each orbital cavity and part of the nasal cavity.*
The zygomatic bone *is better known as the cheekbone.*
The maxilla *is the site of attachment of the upper teeth. It forms part of the hard palate and extends upward, alongside the nose.*
The foramen magnum—*Latin for "large opening"— is the hole in the base of the occipital bone through which the spinal cord passes before it joins the brain.*
Mandible

The vertebral column

Also known as the spinal column or backbone, the vertebral column supports the upper body and head and provides points of attachment for the ribs, pelvis, and many muscles. One of its vital roles is to protect the delicate spinal cord.

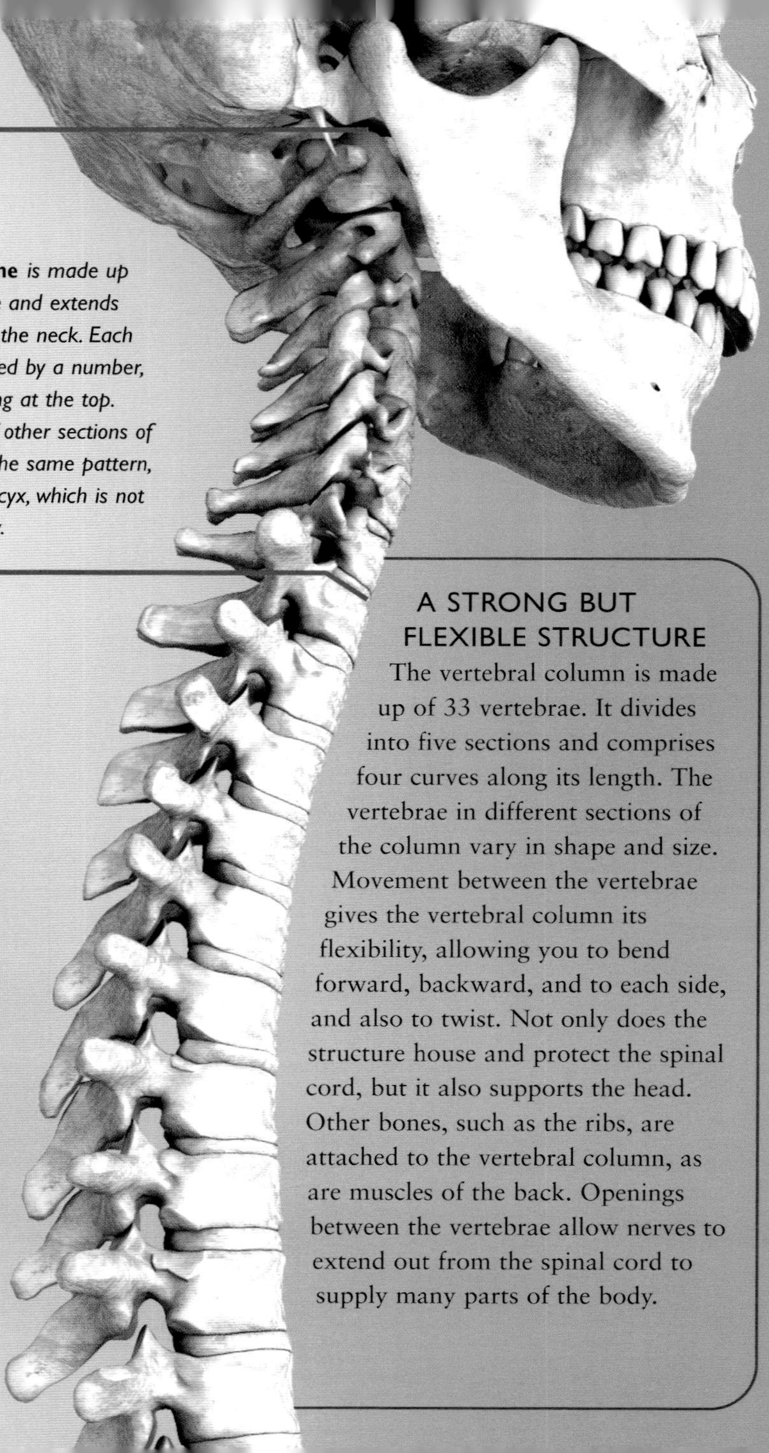

The cervical spine *is made up of seven vertebrae and extends down the back of the neck. Each vertebra is identified by a number, C1 to C7, beginning at the top. The numbering of other sections of the spine follows the same pattern, except for the coccyx, which is not labeled in this way.*

The thoracic spine, *which comprises 12 vertebral bones—T1 to T12—lies at the back of the chest. This section of the spinal column attaches to the ribs.*

A STRONG BUT FLEXIBLE STRUCTURE

The vertebral column is made up of 33 vertebrae. It divides into five sections and comprises four curves along its length. The vertebrae in different sections of the column vary in shape and size. Movement between the vertebrae gives the vertebral column its flexibility, allowing you to bend forward, backward, and to each side, and also to twist. Not only does the structure house and protect the spinal cord, but it also supports the head. Other bones, such as the ribs, are attached to the vertebral column, as are muscles of the back. Openings between the vertebrae allow nerves to extend out from the spinal cord to supply many parts of the body.

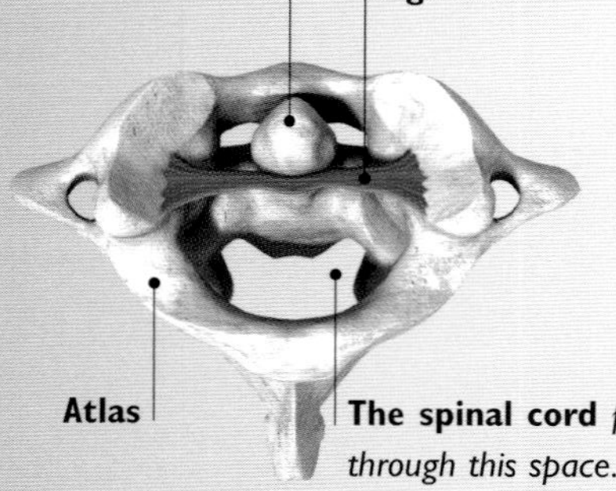
Axis
Ligament
Atlas
The spinal cord passes through this space.

The atlas and axis cervical vertebrae
The top two cervical vertebrae differ significantly from the vertebrae below them. This is because they play a pivotal role in supporting the head and allowing it to move. The first vertebra (C1), called the atlas, is basically a ring of bone that forms two joints with the occipital bone at the back of the skull; this enables you to nod. The second cervical vertebra (C2), called the axis, has a section—the dens—that projects upward through the atlas; this joint enables you to shake your head from side to side.

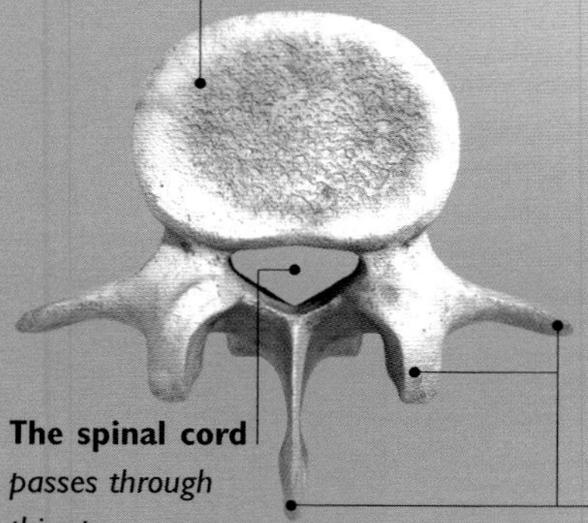
The body of the vertebra is the part that takes the greatest weight and is protected by an intervertebral disc.
The spinal cord passes through this space.
The bony knobs called processes that extend from each vertebra are points of attachment for muscles or between one vertebra and the next.

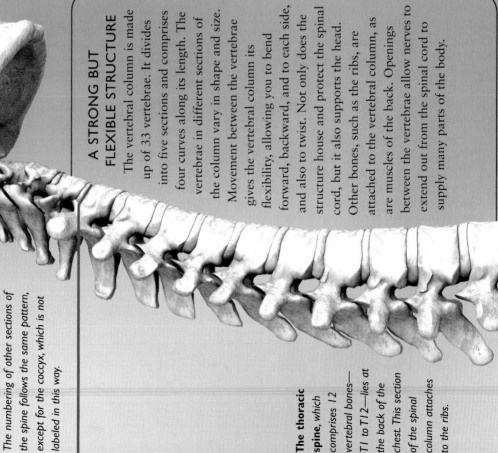

A typical lumbar vertebra

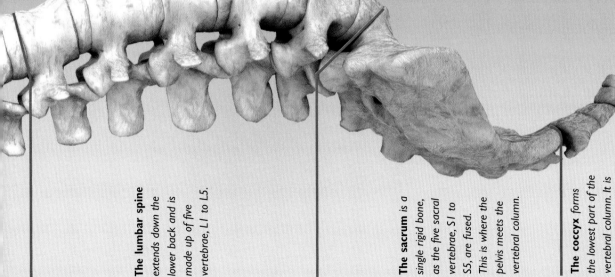

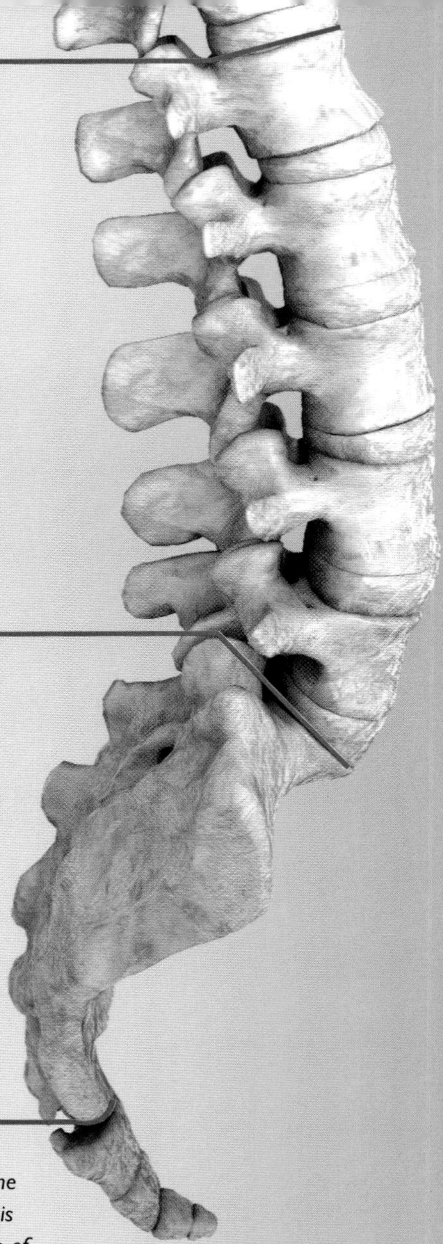
The lumbar spine *extends down the lower back and is made up of five vertebrae, L1 to L5.*
The sacrum *is a single rigid bone, as the five sacral vertebrae, S1 to S5, are fused. This is where the pelvis meets the vertebral column.*
The coccyx *forms the lowest part of the vertebral column. It is rigid, being made up of four fused vertebrae.*

Shock absorbers

Discs lie between the vertebrae and act as shock absorbers; they also form part of the joints between the vertebrae. They are made up of a tough outer layer (the annulus fibrosus) encasing a softer center, known as the nucleus pulposus. Ligaments around and within the vertebral column help keep the discs and the vertebrae in place.

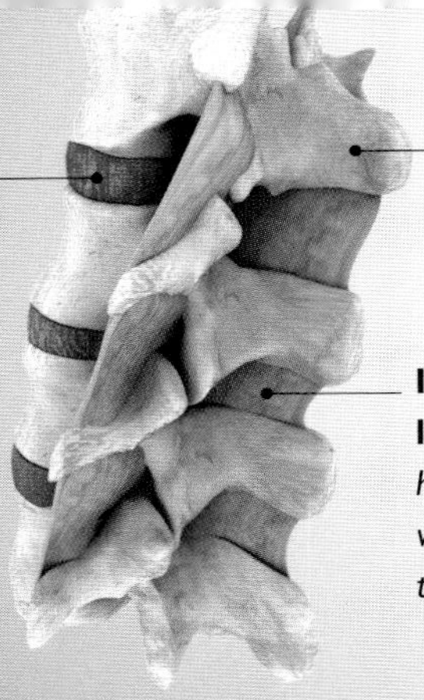
Each intervertebral disc *has a fibrous exterior but a gelatinous center for maximum cushioning effect.*
A bony vertebra
Interspinous ligaments *help hold the vertebrae together.*

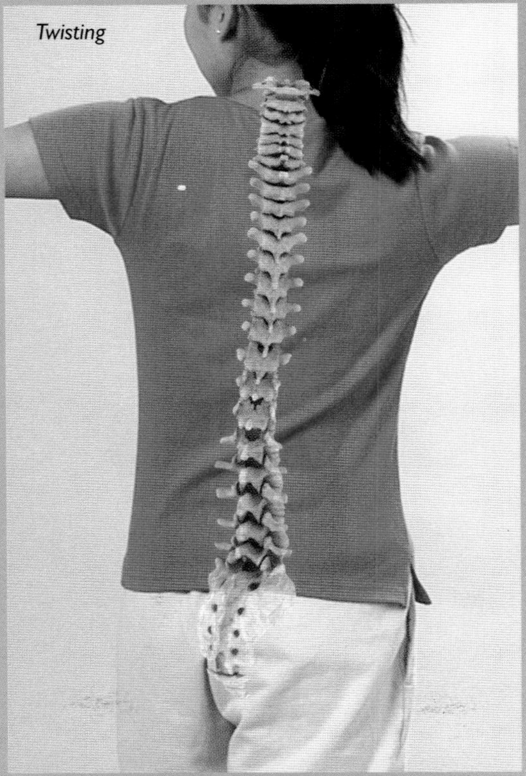
Twisting

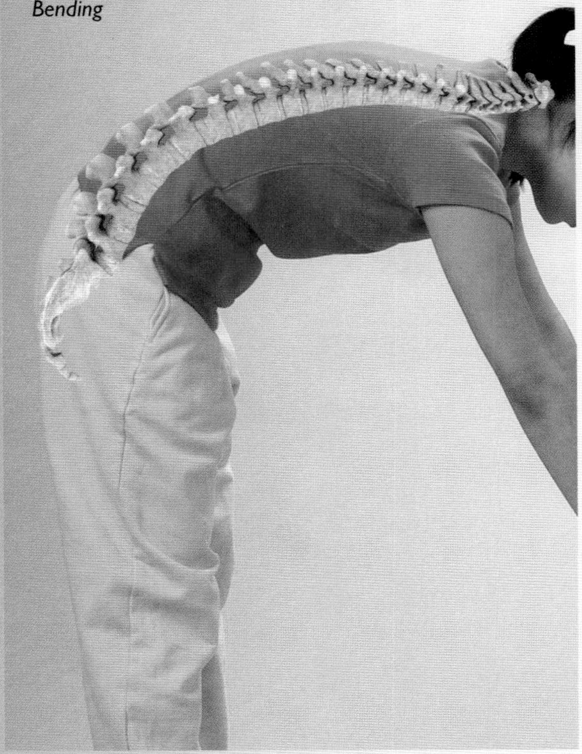
Bending

The ingenious interlocking structure *of the vertebrae, together with the cushioning discs and supporting ligaments, enables the spine to bend and twist as required.*

The structure of bone

Bones have remarkable strength and yet are light enough to allow us to move with great agility and speed. Their structure also means they can give slightly, enabling them to withstand jolts when we move.

WHAT MAKES A BONE?

Most bones are made up of a hard outer layer (compact bone) encasing a less dense inner part (cancellous bone). Long bones also have a central core of bone marrow. Bone is made of living cells, protein fibers, minerals, and water. These components are arranged differently in the two types of bone.

- Compact bone appears solid to the naked eye. However, it consists of numerous osteons, each comprising concentric rings of bone (lamellae) lying around a central canal, which contains vessels and nerves. Between the bony layers are tiny gaps called lacunae, each of which is connected to the central canal.
- Cancellous bone looks like a honeycomb. It is made up of numerous rigid struts called trabeculae that form an interconnected framework.

The periosteum—*shown here as if peeled back—is a fibrous membrane that surrounds all bones. It contains the blood vessels that supply the bone beneath with nutrients.*

Compact bone *is made up of cylindrical units called osteons, also known as Haversian systems. A typical osteon measures a little less than 0.05 inch in diameter. The densely packed arrangement of osteons within compact bone gives this type of bone its great strength.*

In the lacunae—*the minute gaps between the lamellae—are fluid and mature bone cells called osteocytes.*

Lamellae *are concentric growth rings of calcified tissue that surround the central canal of each osteon.*

A central canal *runs through the middle of each osteon. It contains blood vessels and (not shown here) nerves and lymph vessels.*

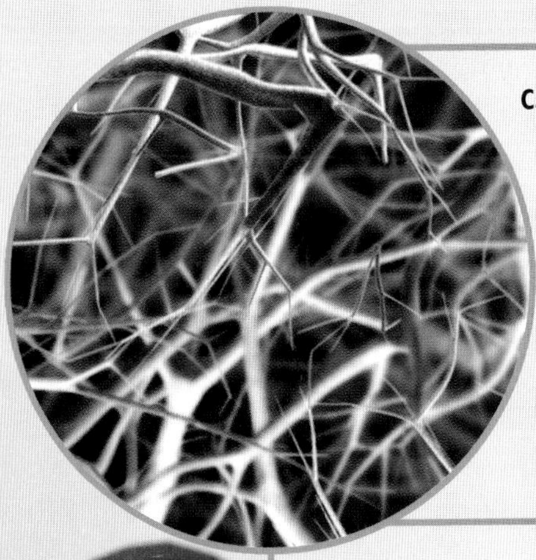

Cancellous bone *consists of narrow bars of bony tissue called trabeculae (the word trabecula means a small supporting beam or bar in Latin). The trabeculae provide a framework of rigid struts, with numerous tiny interconnecting spaces between them. Within these spaces lies the bone marrow.*

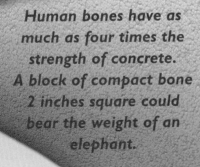

Human bones have as much as four times the strength of concrete. A block of compact bone 2 inches square could bear the weight of an elephant.

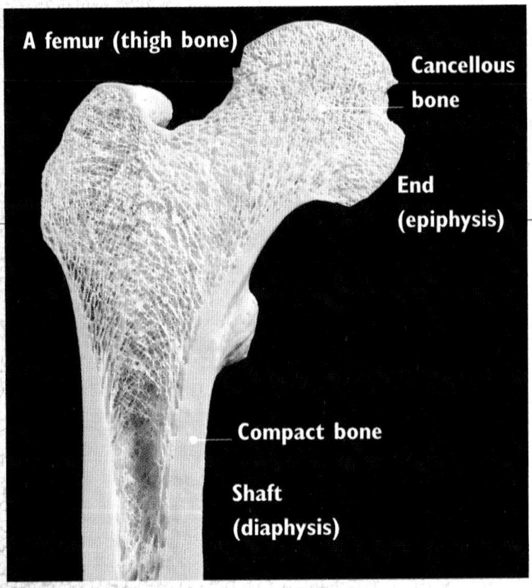
A femur (thigh bone)
Cancellous bone
End (epiphysis)
Compact bone
Shaft (diaphysis)

A look at a long bone

A long bone is made up of a shaft (the diaphysis) and two ends (called the epiphyses).

- The shaft consists of a thick outer layer of compact bone encasing cancellous bone, within which there is a central cavity. Both cancellous bone and central cavity contain bone marrow.
- Each end of the bone consists mainly of cancellous bone with a thin outer layer of compact bone.

How bones are made

A baby is born with many bones partly consisting of bendable cartilage. In childhood and adolescence, as these bones grow in size, the cartilage is gradually replaced by much harder bone tissue.

STARTING AT THE BEGINNING

Long, short, and irregular bones all develop from rods of cartilage within the fetus. For example, each long bone begins in the fetus as a length of cartilage that gradually hardens—a process called ossification—in the middle of its shaft. Ossification is the result of the work of osteoblasts. These cells produce osteoid, the substance that when calcified, becomes bone. At the same time, the bone slowly grows in size.
After the baby is born, each long bone continues to lengthen and harden, but the centers of growth and ossification are now at the ends of the bone. Ossification and bone growth finally ends when you reach age 20 or so.

Flat bones (as in the skull) and sesamoid bones (such as the kneecap) develop slightly differently. Flat bones develop from membranous tissue rather than cartilage, and sesamoid bones develop from tendon-type tissue.

LONG BONE OF A NEWBORN BABY

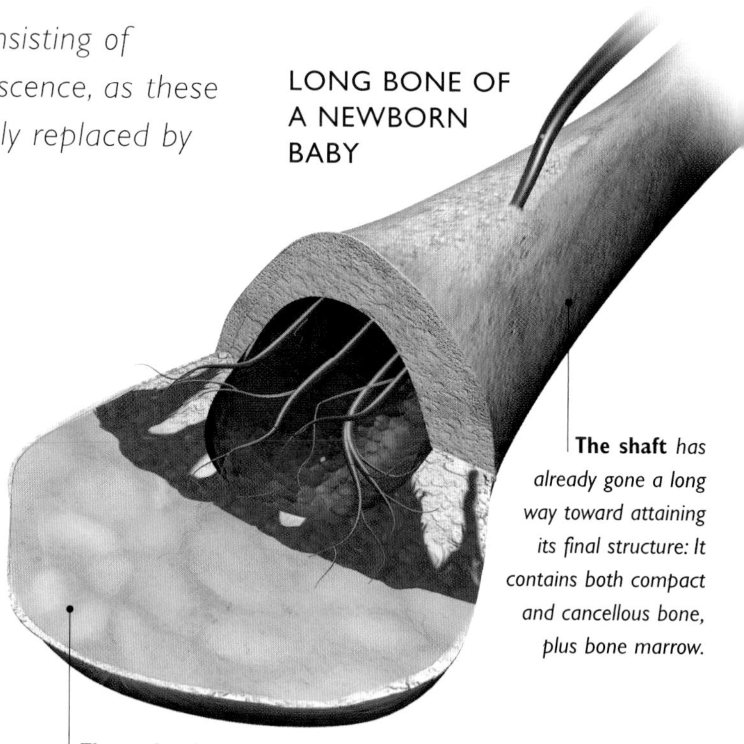

The shaft *has already gone a long way toward attaining its final structure: It contains both compact and cancellous bone, plus bone marrow.*

The ends of each long bone *are still cartilage (soft elastic connective tissue) because ossification has yet to take place. This allows the bone to grow.*

HOW A BROKEN BONE HEALS ITSELF

The body has a remarkable capacity for repair. This is clearly demonstrated by the way bone fractures heal. For the healing process to be effective, the ends of the broken bone must first be brought together and then held in place.

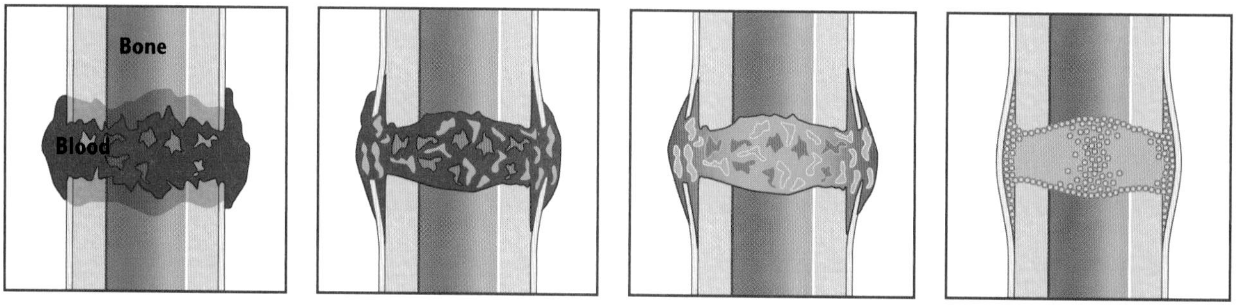
Bone
Blood

Blood vessels in the bone *are severed when a bone is broken. The blood that accumulates at the site of the fracture begins to clot, sealing off the damaged blood vessels.*

Fibrous healing tissue *gradually replaces the clotted blood and holds the broken ends together. New blood vessels develop to supply oxygen and nutrients to the fibrous tissue.*

Osteoblasts *move into the fibrous tissue and begin the process of ossification. The fibrous tissue begins to mineralize and harden. At this stage, it is called a callus.*

As ossification proceeds, *the bone returns as closely as possible to its original shape and structure. The bone formation process may take four to six weeks, with total recovery taking several months.*

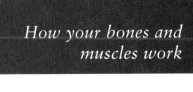

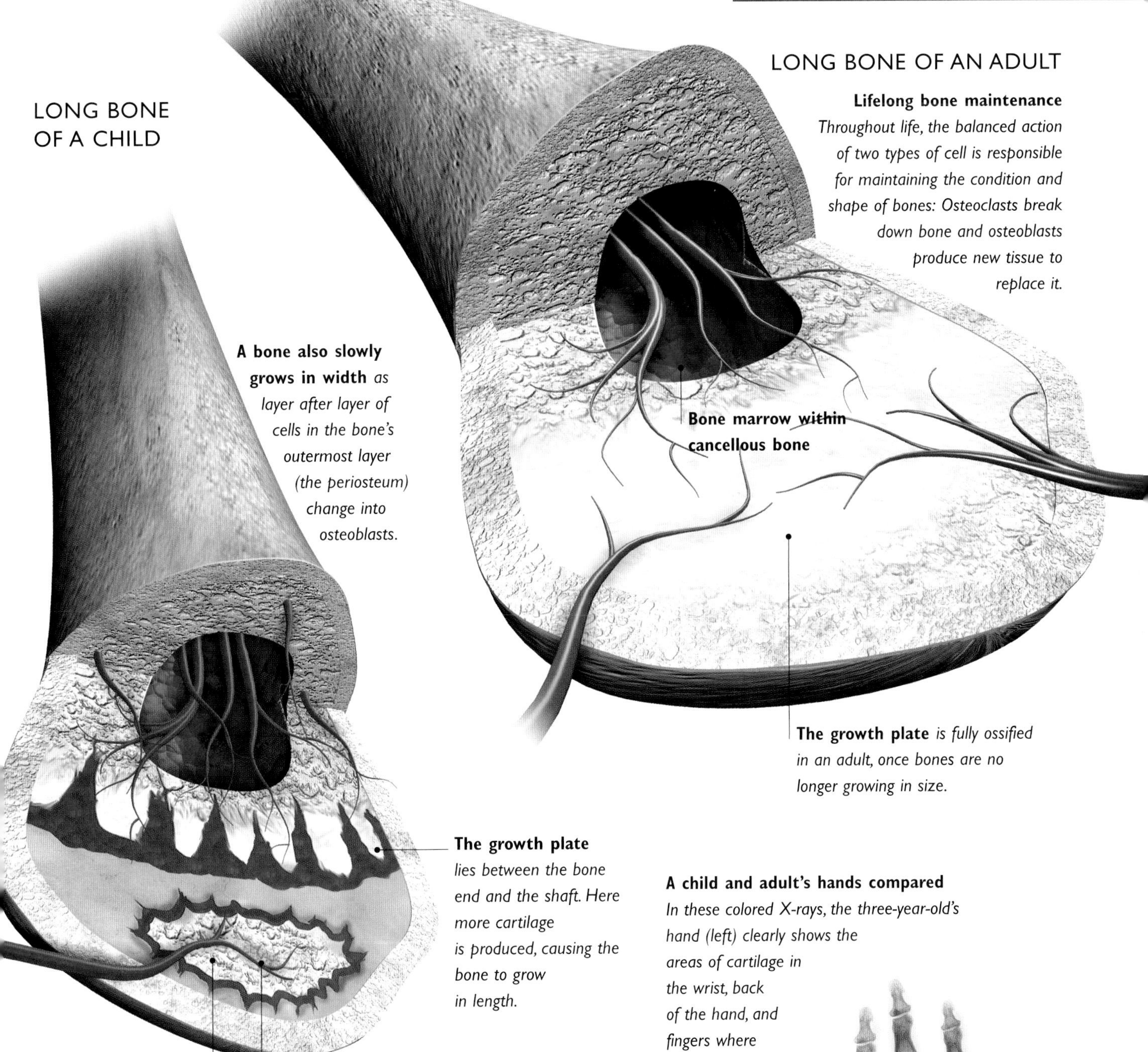

LONG BONE OF AN ADULT

Lifelong bone maintenance
Throughout life, the balanced action of two types of cell is responsible for maintaining the condition and shape of bones: Osteoclasts break down bone and osteoblasts produce new tissue to replace it.

Bone marrow within cancellous bone

The growth plate *is fully ossified in an adult, once bones are no longer growing in size.*

LONG BONE OF A CHILD

A bone also slowly grows in width *as layer after layer of cells in the bone's outermost layer (the periosteum) change into osteoblasts.*

The growth plate *lies between the bone end and the shaft. Here more cartilage is produced, causing the bone to grow in length.*

The site of ossification *is where the cartilage is gradually converted into cancellous bone.*

Blood vessels *supply the tissue of the ossification site with oxygen and nutrients.*

A child and adult's hands compared
In these colored X-rays, the three-year-old's hand (left) clearly shows the areas of cartilage in the wrist, back of the hand, and fingers where ossification has not yet taken place.

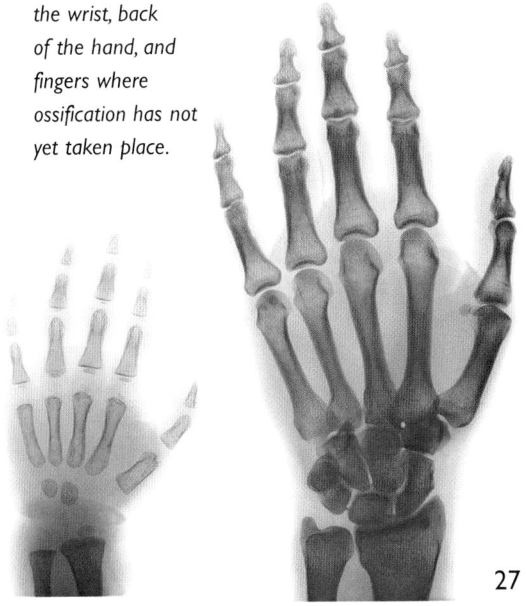

At birth, you had more than 300 largely cartilaginous bones. As these hardened, the number decreased to 206 because some smaller bones fused together.

Joints—making connections

You can move easily despite the rigidity of your bones because bones are linked to each other at intersections called joints. Not all joints allow bones to move freely, however. Some are fixed and others allow only slight movement.

A joint is "double-jointed" when the ligaments are longer than normal, allowing the joint a greater range of movement than usual.

TYPES OF JOINT

- **Movable joints** Joints that allow a fair amount of movement are called synovial joints. They include hinge joints, ball-and-socket joints, and pivot joints: Types vary according to the shape of the bones involved and the movements they allow the bones to make.
- **Fixed joints** Two bones connected by a fixed joint are held together with fibrous collagen that allows almost no movement. Fixed joints occur where movement of one bone on another is not wanted, such as between the bones making up the cranium.
- **Joints that move a little** The ends of adjoining bones in joints that allow only a small amount of movement are covered by slippery hyaline cartilage, then connected by tougher fibrocartilage. Examples include the joints between the vertebrae and the joint where the two sides of the pelvis meet at the front (the symphysis pubis).

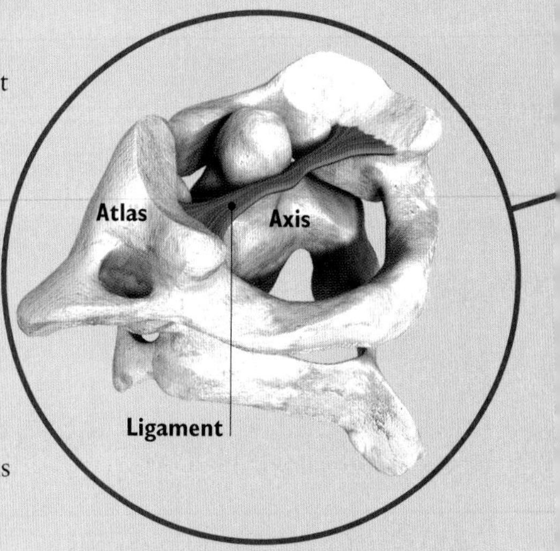

Atlas
Axis
Ligament

Synovial pivot joint
The first vertebra of the spinal column, called the atlas, and the second vertebra, the axis, meet at a pivot joint. This allows the head to turn from side to side.

Supporting and stabilizing synovial joints

- **The muscles** surrounding a joint are its main supports. The muscle holds the joint ends together and brings them back into alignment after movement.
- **Ligaments** are fibrous bands that bind two bones together at a joint. They are made up of dense, flexible but inelastic connective tissue. A ligament helps stabilize a joint by preventing excessive movement; once taut, the ligament cannot be stretched further.
- **A capsule of tough fibrous tissue** encases each joint. In addition to protecting the joint, it keeps the bones in place.

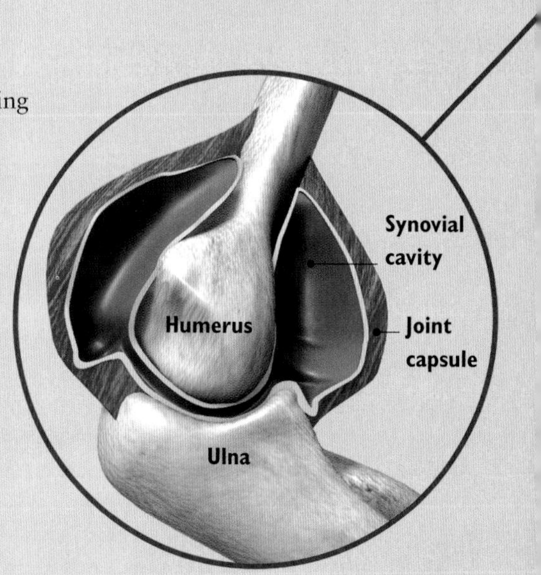

Synovial cavity
Humerus
Joint capsule
Ulna

Synovial hinge joint
The elbow is a hinge joint, which can bend and straighten. The knee and ankle are also hinge joints, as are joints in the fingers and toes and the joint where the first vertebra (the atlas) is connected to the occipital bone of the skull, allowing you to nod your head up and down.

FEATURES SHARED BY ALL SYNOVIAL JOINTS

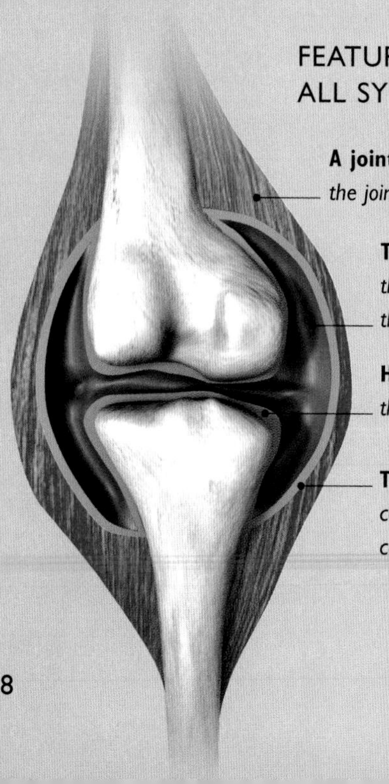

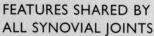

A joint capsule *surrounds and protects the joint. It is made of fibrous tissue.*

The synovial cavity *is the space around the bone ends. It contains synovial fluid that lubricates all joint movements.*

Hyaline cartilage *covers all joint surfaces that are in contact with other bone.*

The synovial membrane *lines the joint capsule and any bony surfaces not covered by hyaline cartilage.*

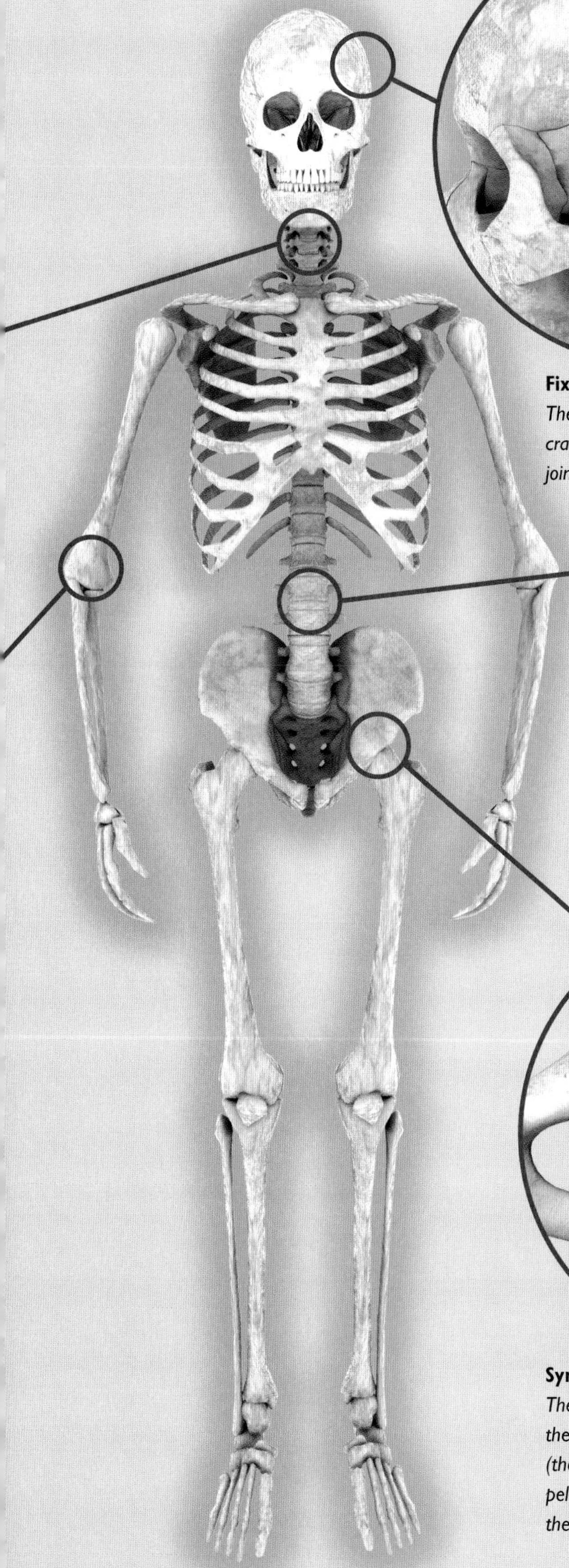

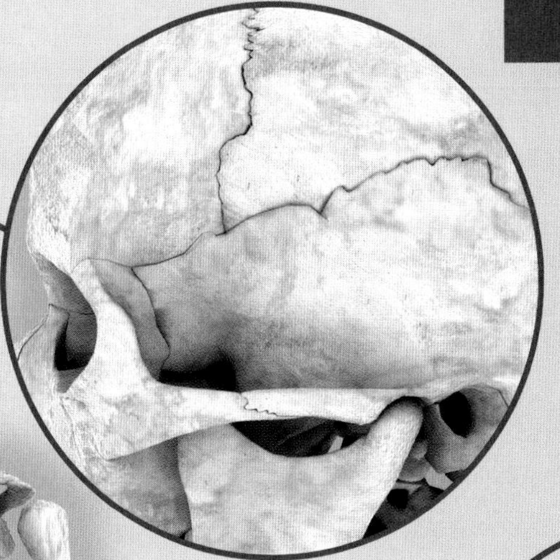

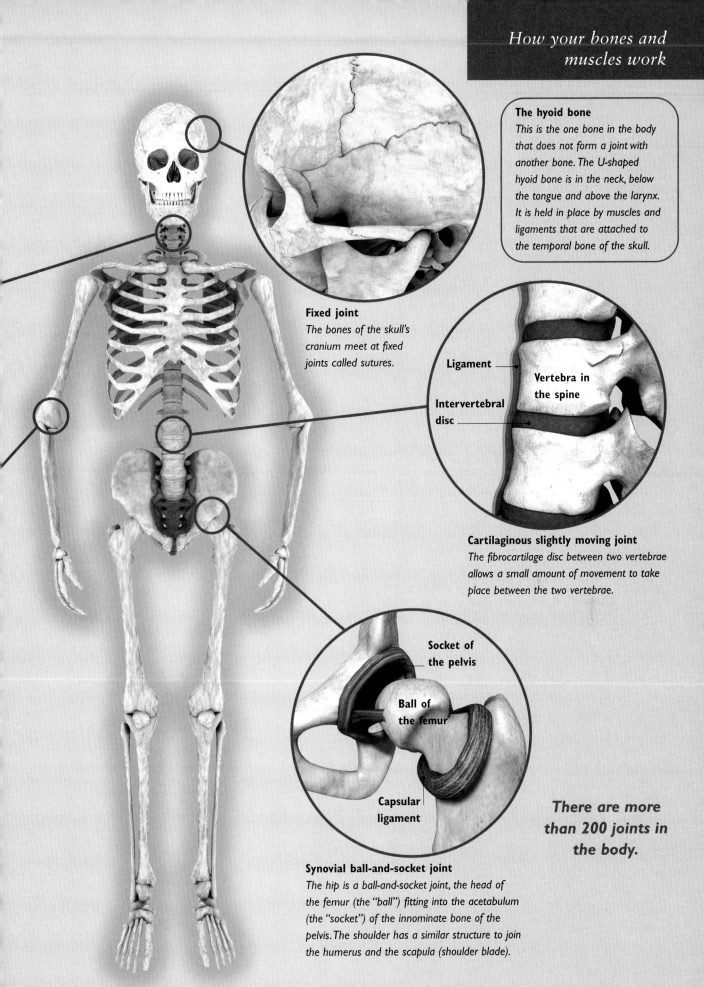

The hyoid bone

This is the one bone in the body that does not form a joint with another bone. The U-shaped hyoid bone is in the neck, below the tongue and above the larynx. It is held in place by muscles and ligaments that are attached to the temporal bone of the skull.

Fixed joint

The bones of the skull's cranium meet at fixed joints called sutures.

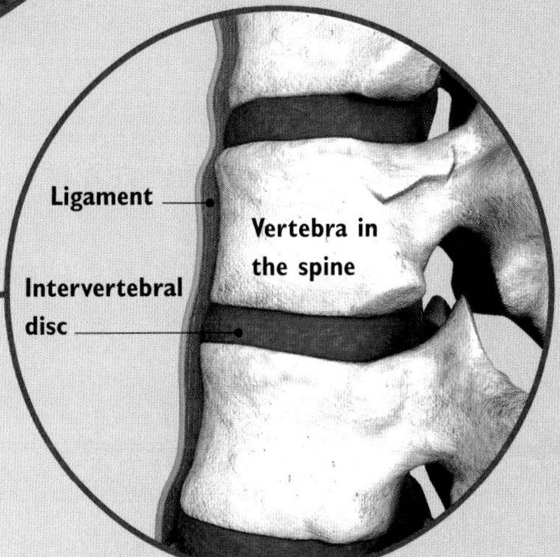
Ligament
Vertebra in the spine
Intervertebral disc

Cartilaginous slightly moving joint

The fibrocartilage disc between two vertebrae allows a small amount of movement to take place between the two vertebrae.

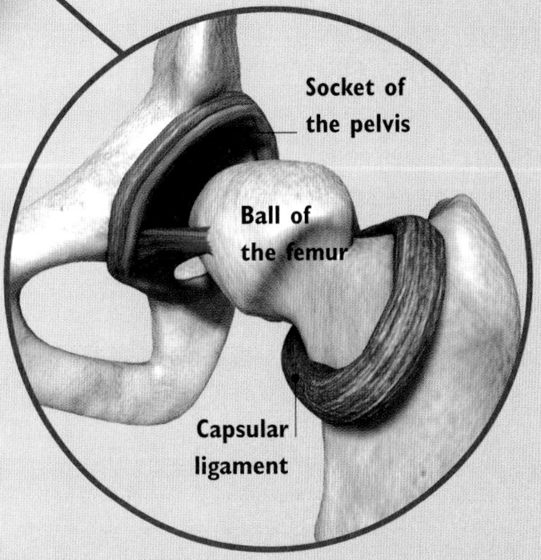
Socket of the pelvis
Ball of the femur
Capsular ligament

Synovial ball-and-socket joint

The hip is a ball-and-socket joint, the head of the femur (the "ball") fitting into the acetabulum (the "socket") of the innominate bone of the pelvis. The shoulder has a similar structure to join the humerus and the scapula (shoulder blade).

There are more than 200 joints in the body.

The skeletal muscles

Muscle is tissue that can contract and thus produce movement. As well as moving and steering the various parts of the skeleton, the skeletal muscles help maintain your position and posture when you are motionless by remaining partially contracted at all times.

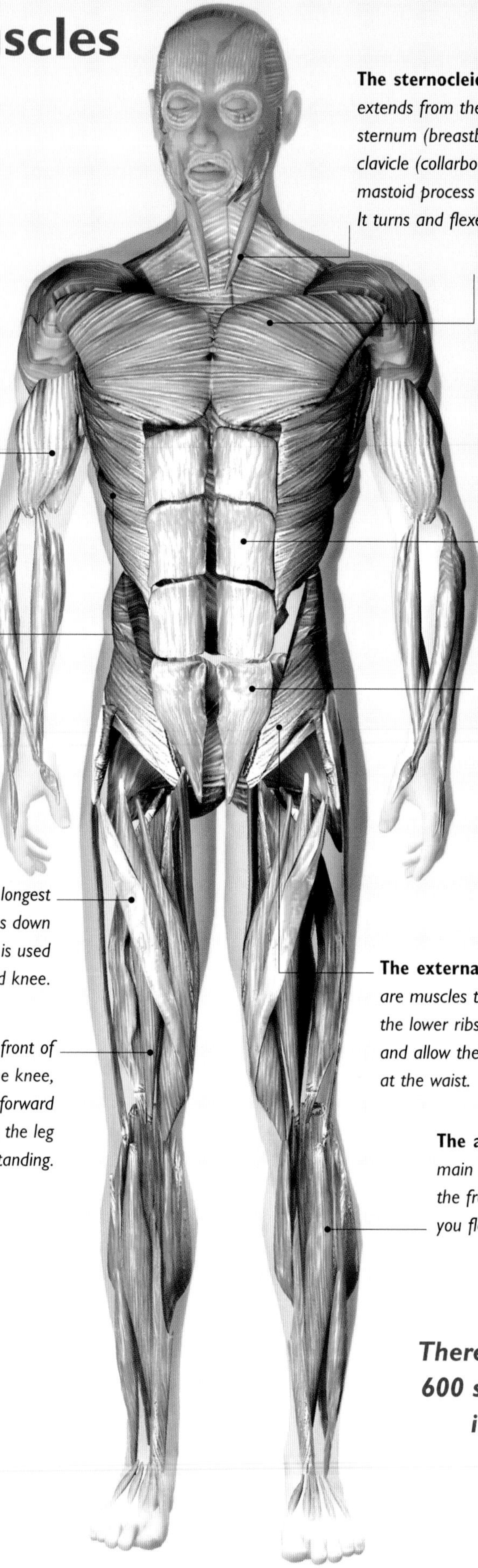
The sternocleidomastoid *extends from the top of the sternum (breastbone) and clavicle (collarbone) to the mastoid process behind the ear. It turns and flexes the neck.*
The pectoralis major *is a large fan-shaped muscle on the chest; it acts with the latissimus dorsi to pull the arm down or pull the body up, as when climbing.*
The biceps *at the front of the upper arm contracts to bend the elbow and the arm. It works in tandem with the triceps.*
The abdominal muscles (abs) *contract to bend the body forward.*
The serratus anterior *is situated between the ribs and the shoulder blade; it draws the shoulder blade forward, such as when you reach out in front of you.*
The rectus abdominus *extends along the length of the front of the abdomen; it enables the trunk to bend forward and sideways. ("Rectus" means straight.)*
The sartorius *is the longest muscle in the body. It runs down the front of the thigh and is used to flex the hip and knee.*
The external obliques *are muscles that run from the lower ribs to the pelvis and allow the body to twist at the waist.*
The quadriceps *at the front of the thigh straightens the knee, pulling the lower leg forward when walking and holding the leg straight when standing.*
The anterior tibialis *is the main muscle that runs down the front of the calf. It helps you flex your foot.*

UNDER CONTROL

Skeletal muscles vary from tiny muscles in the hands or feet to the huge muscles of the buttocks and thighs. The actions of skeletal muscles are under our conscious control, although many of the movements we perform occur without a second thought.

There are more than 600 skeletal muscles in the body.

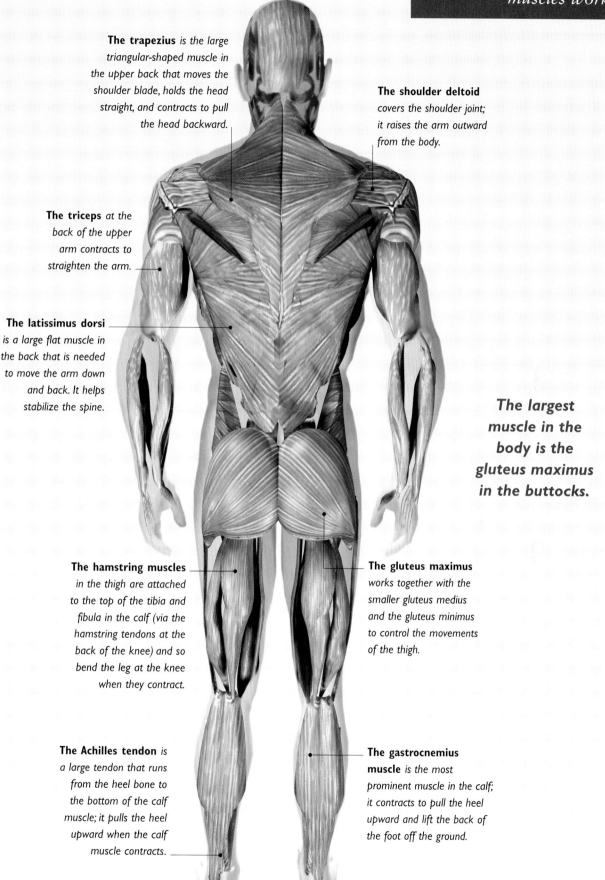

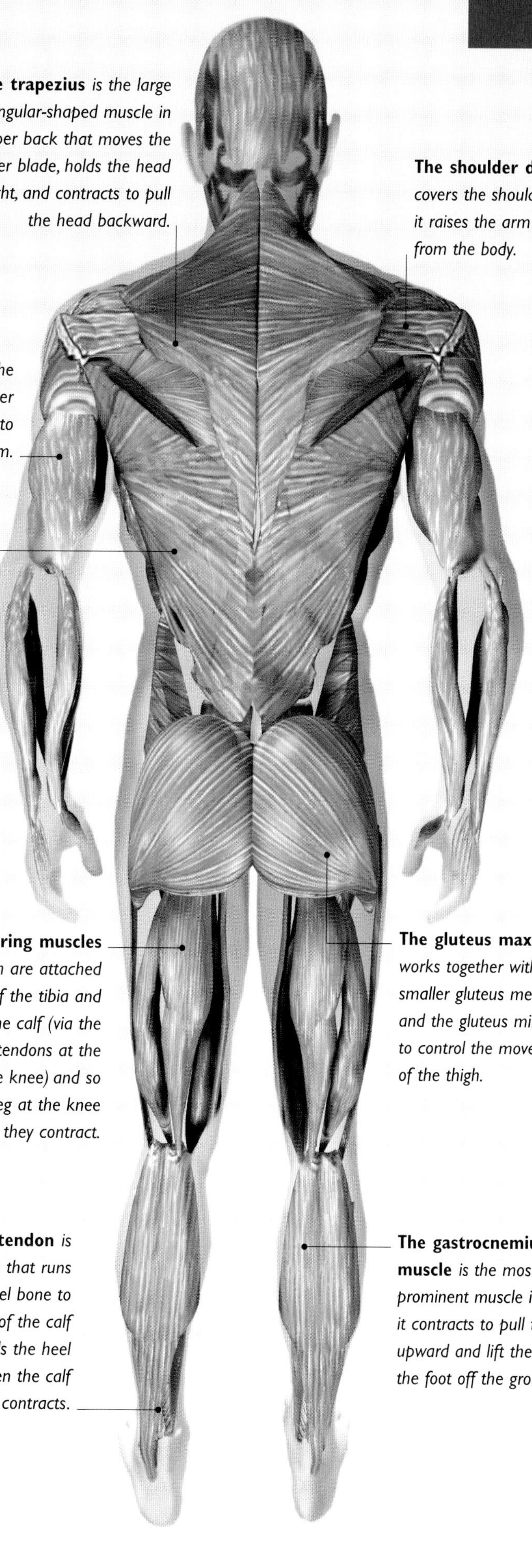
The trapezius *is the large triangular-shaped muscle in the upper back that moves the shoulder blade, holds the head straight, and contracts to pull the head backward.*
The shoulder deltoid *covers the shoulder joint; it raises the arm outward from the body.*
The triceps *at the back of the upper arm contracts to straighten the arm.*
The latissimus dorsi *is a large flat muscle in the back that is needed to move the arm down and back. It helps stabilize the spine.*
The hamstring muscles *in the thigh are attached to the top of the tibia and fibula in the calf (via the hamstring tendons at the back of the knee) and so bend the leg at the knee when they contract.*
The gluteus maximus *works together with the smaller gluteus medius and the gluteus minimus to control the movements of the thigh.*
The Achilles tendon *is a large tendon that runs from the heel bone to the bottom of the calf muscle; it pulls the heel upward when the calf muscle contracts.*
The gastrocnemius muscle *is the most prominent muscle in the calf; it contracts to pull the heel upward and lift the back of the foot off the ground.*

The largest muscle in the body is the gluteus maximus in the buttocks.

The structure of skeletal muscle

Skeletal muscles are made up of long fibers—some as long as 12 inches—that lie next to each other in parallel bundles. Very rapid powerful contractions of these individual fibers cause the muscles to move.

Muscle fiber makes up approximately two fifths of the average person's body weight.

A CLOSER LOOK

The bundles of long fibers within skeletal muscles are known as fascicles. The fibers themselves consist of tinier fibers called myofibrils, which in turn are made of even thinner fibers called myofilaments. There are two types of myofilament, thick and thin, and it is the arrangement of these two types that gives muscle fibers their characteristic striped appearance. Myofilaments slide between each other to shorten the muscle and so produce muscle contraction. This action is triggered by nerve impulses.

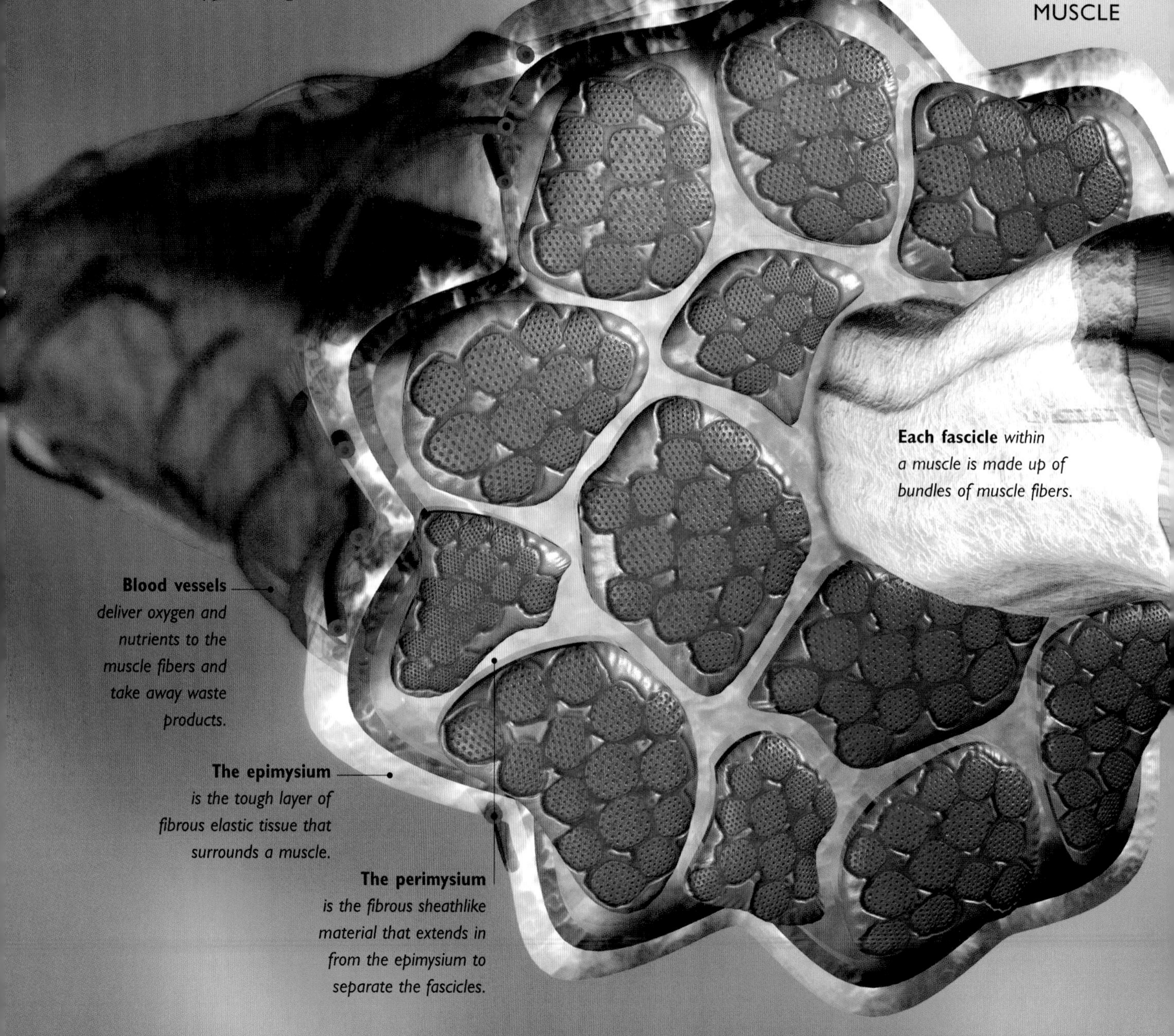
CROSS SECTION OF A SKELETAL MUSCLE
Each fascicle *within a muscle is made up of bundles of muscle fibers.*
Blood vessels *deliver oxygen and nutrients to the muscle fibers and take away waste products.*
The epimysium *is the tough layer of fibrous elastic tissue that surrounds a muscle.*
The perimysium *is the fibrous sheathlike material that extends in from the epimysium to separate the fascicles.*

UNDER THE MICROSCOPE—THE THREE TYPES OF MUSCLE IN THE BODY

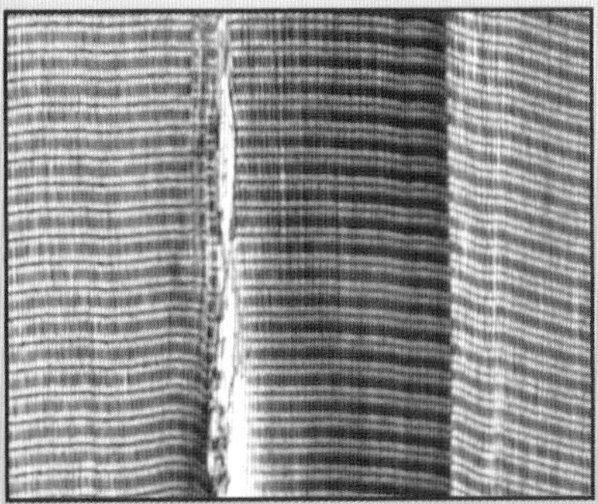

Skeletal or striated muscle
These are the muscles that are attached to the bones of the skeleton. They are sometimes called voluntary muscles, because they are under our control, or striated (striped) muscle, because of the alternating dark and light bands seen on the muscle fibers when they are viewed under a microscope.

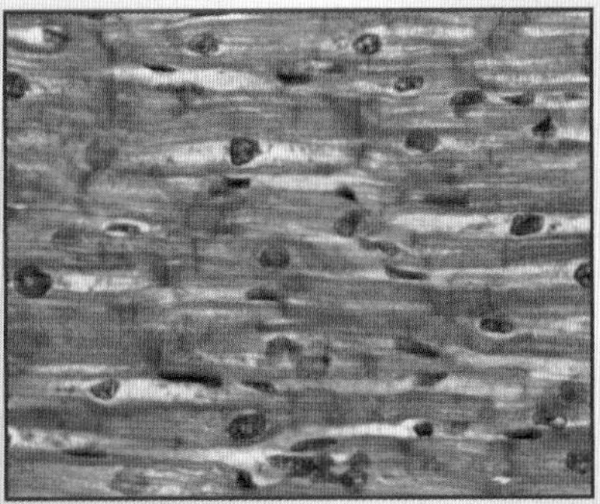

Smooth muscle
Smooth muscles are located in the walls of blood vessels and hollow organs such as the stomach and GI tract. Single spindle-shaped cells held together by connective tissue contract and relax slowly and rhythmically to push their contents forward. These contractions are not under voluntary control.

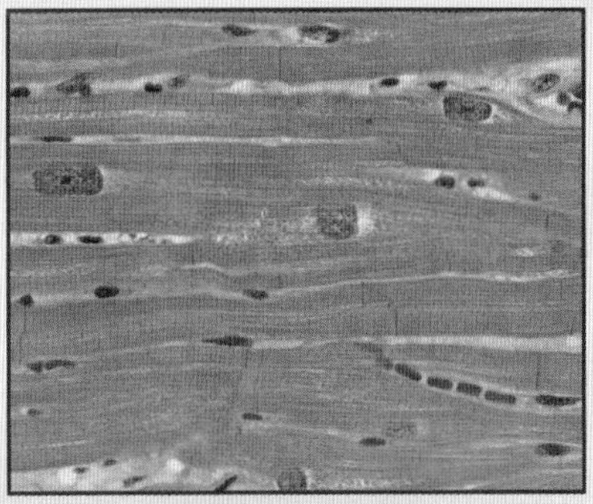

Cardiac or heart muscle
Cardiac muscle is found only in the heart wall. It makes rapid rhythmic contractions that pump blood around the body. These contractions are not under conscious control. Each cardiac muscle cell consists of one or two elongated nuclei and an extensive branching cytoplasm.

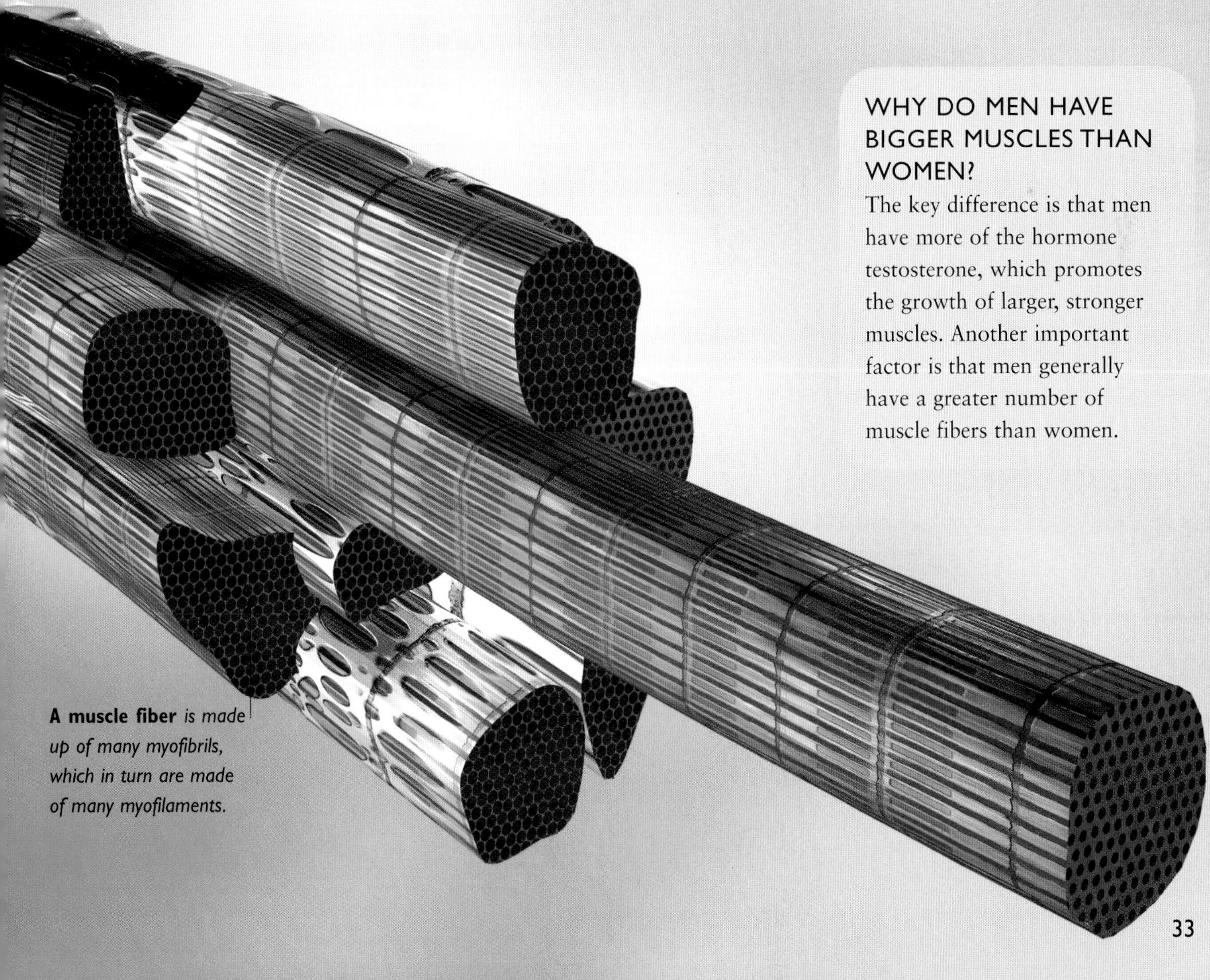

A muscle fiber *is made up of many myofibrils, which in turn are made of many myofilaments.*

WHY DO MEN HAVE BIGGER MUSCLES THAN WOMEN?

The key difference is that men have more of the hormone testosterone, which promotes the growth of larger, stronger muscles. Another important factor is that men generally have a greater number of muscle fibers than women.

Making movements

Bones and muscles make movement possible by working together. The muscles, stimulated and synchronized by nerve impulses originating in the brain and spinal cord, provide the mechanical power to move the bones.

The human body uses more than 200 separate muscles just to take a few steps.

HOW THE SKELETAL MUSCLES WORK

Skeletal muscles extend across all the joints in the skeleton, and when they contract they either pull bones together or pull them apart, depending on where they are attached to the bones. The muscles are arranged in opposing pairs or groups: Some contract while others relax, enabling movement to work in more than one direction. A classic example is the biceps and triceps pairing in the upper arm. To raise the forearm, the biceps at the front of the upper arm contracts and shortens while the triceps at the back relaxes and lengthens. To reverse the action and lower the forearm, the actions of the muscles are reversed—the biceps relaxes while the triceps contracts.

CONNECTING WITH THE BRAIN

For every movement you make, your brain is processing information to help you on your way. Sensory nerves in a tendon (the tough, fibrous cord, made of collagen, that attaches muscle to bone) set off a reflex contraction in the attached muscle. Within muscles and tendons, specialized nerve endings called proprioceptors monitor the degree of muscle contraction so that the brain can make adjustments as necessary when performing a task. The brain uses this feedback, together with information from the ears and eyes, to maintain the body's balance and posture. Proprioceptors also tell your brain where the parts of your body are in relation to each other and in relation to the things around you.

BEST FOOT FORWARD—TAKING A STEP

1 Heel strikes ground

The calf muscles pull the long tendons that connect them to the bones of the ankle and foot. This causes the ankle to flex and points the toes upward (dorsiflexion), so that the heel bone—the largest of the seven tarsal bones—hits the ground first and absorbs the shock of the impact.

2 In mid step

Your midfoot, made up of the smaller tarsals and the metatarsal bones, forms an arch that bears the weight of your body as it rolls along the length of the foot from your heel to your big toe. The bones are bound together by strong ligaments that allow movement and flexibility while maintaining a firm base. This is particularly important when walking over uneven ground.

3 Pushing off

The calf muscles contract and pull on the Achilles tendon, lifting the heel from the ground. At the same time, the tendons that extend along the length of the upper foot contract, flexing the joints between the metatarsal bones and the phalanges in the toes, ready to push off from the forefoot.

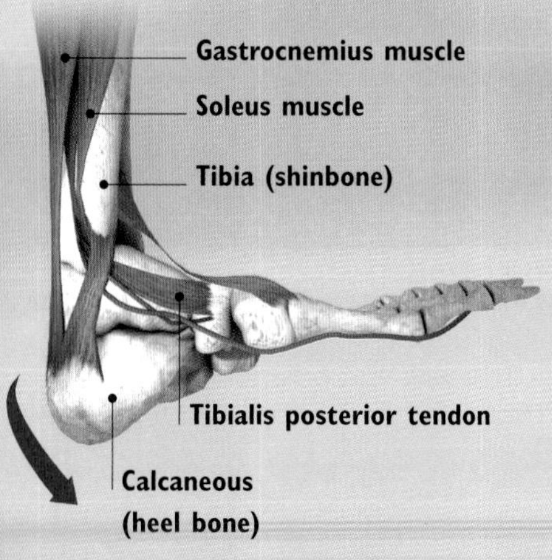
Gastrocnemius muscle
Soleus muscle
Tibia (shinbone)
Tibialis posterior tendon
Calcaneous (heel bone)

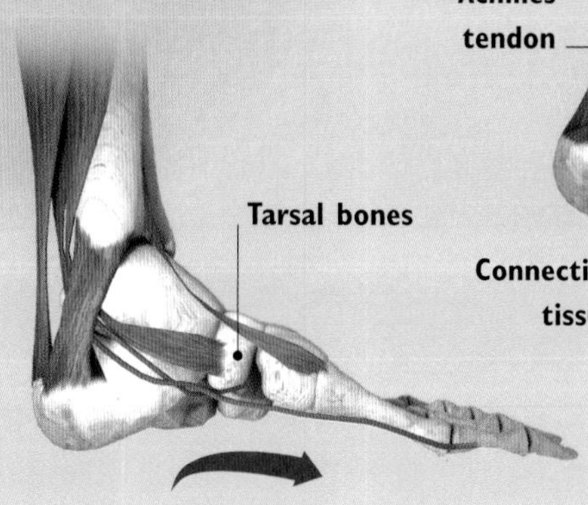
Tarsal bones

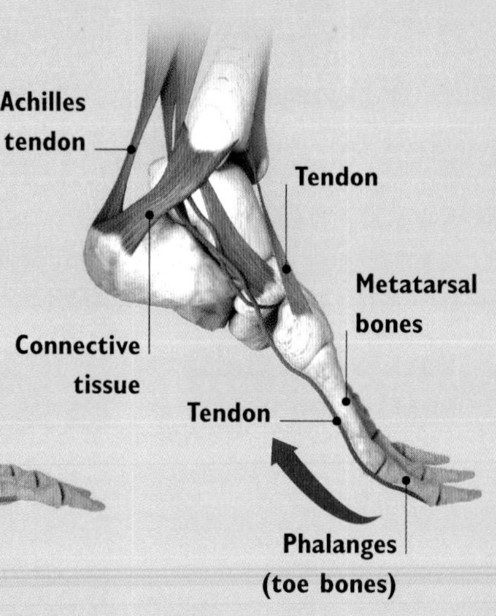
Achilles tendon
Tendon
Metatarsal bones
Connective tissue
Tendon
Phalanges (toe bones)

When you walk

In walking to work, as in all other activities, your movements are smooth and steady. This is thanks to the cerebellum in the hind part of the brain. The cerebellum coordinates the many automatic movements made by skeletal muscles all over the body in order to maintain balance and stability. Clusters of nerve tissue deep within the brain called basal ganglia also help to keep movements smooth by coordinating muscle activity.

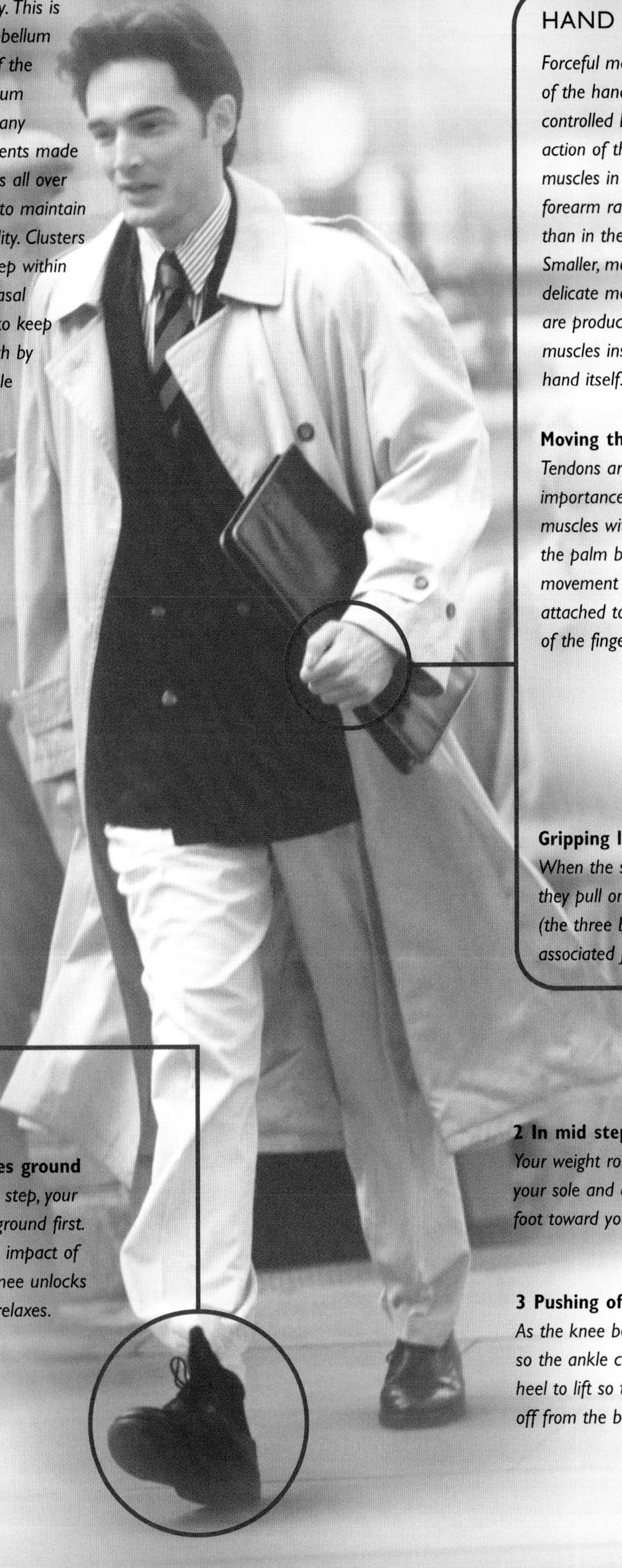

HAND CONTROL

Forceful movements of the hand are controlled by the action of the muscles in the forearm rather than in the hand. Smaller, more delicate movements are produced by the muscles inside the hand itself.

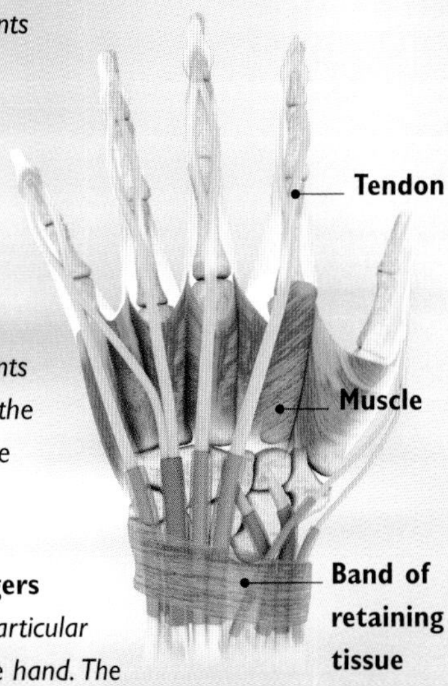
Tendon
Muscle
Band of retaining tissue

Moving the fingers

Tendons are of particular importance in the hand. The muscles within the forearm and in the palm both produce finger movement via the long tendons attached to the bones of the fingers.

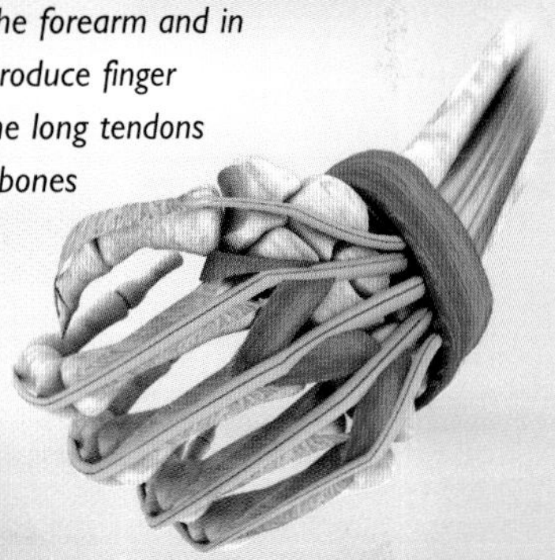

Gripping lightly with the fingers

When the short muscles in the palm of the hand contract, they pull on the tendons attached to the three phalanges (the three bones in each finger) and cause the three associated joints to flex and the fingers to curl inward.

1 Heel strikes ground

As you take a step, your heel hits the ground first. To absorb the impact of landing, the knee unlocks and the foot relaxes.

2 In mid step

Your weight rolls down the outside of your sole and across the ball of your foot toward your big toe.

3 Pushing off

As the knee bends forward, the calf contracts so the ankle can also bend. This causes the heel to lift so that you can continue pushing off from the ball of your foot to your big toe.

A day in the life of your bones and muscles

At any time in the day, some of your skeletal muscles will be hard at work, making your bones move or just keeping you upright. All of these muscle contractions are stimulated by nerve impulses originating in the brain and spinal cord.

THE CONTROL CENTER

The brain acts as coordinator of the actions of your muscles, enabling you to perform simple and complex tasks with similar ease. Consider for a moment how many different movements are involved in performing a task as seemingly simple as making a cup of tea. The impulses that activate muscle contractions originate in the motor cortex in the upper part of the brain. However, various other parts of the brain and sensors around the body are also involved in producing every movement that you make.

7:30 A.M. Rise, shine, and stretch

Taking time each morning to limber up, systematically stretching and relaxing muscles and joints to prepare them for the exertions of the day ahead, is the best way to avoid stiffness, strain, and possible injury.

1:00 P.M. Eating for strength

Poached salmon and a green leafy salad make an excellent lunchtime choice for strong bones and healthy muscles. Both are prime sources of calcium, without which bones cannot develop properly, and also contain phosphorus. In addition, salmon has vitamin D and green leafy vegetables provide magnesium.

5:30 P.M. Let's get physical—working those muscles

Time for a trip to the gym. A measured exercise program, over a period of months allows your body to accustom itself to increasing levels of exercise. Greater exercise tolerance is the result of several factors, including an increased capacity of the cardiovascular system and the lungs to meet the muscles' needs for oxygen and nutrients during exercise. Equally important, repeated exercise of specific muscles results in an increase in muscle bulk and strength.

Keeping active
Regular maintenance of your musculoskeletal system—by means of frequent, regular exercise and a healthy diet—greatly increases your chances of remaining fit and active as you grow older.

10:00 P.M. Winding down

Soothing tired muscles with a warm bath makes sense. Water treatments have been valued for their therapeutic effects since ancient times. Soaking in warm water increases the flow of blood to the skin and muscles by enlarging blood vessels close to the skin, thereby warming and relaxing these tissues.

8:30 P.M. A little home repair

Each movement you make requires a complex interplay of different muscles, of which you are usually entirely unaware. You want to drive a screw into a piece of wood, for example. You grip the screwdriver with the flexor muscles of your fingers and turn it clockwise. Simultaneous action of the hand's extensor muscles ensures that the wrist remains straight. The screw is hard to move and you twist harder, mainly with the biceps of the upper arm, and to keep the elbow steady the triceps must also contract.

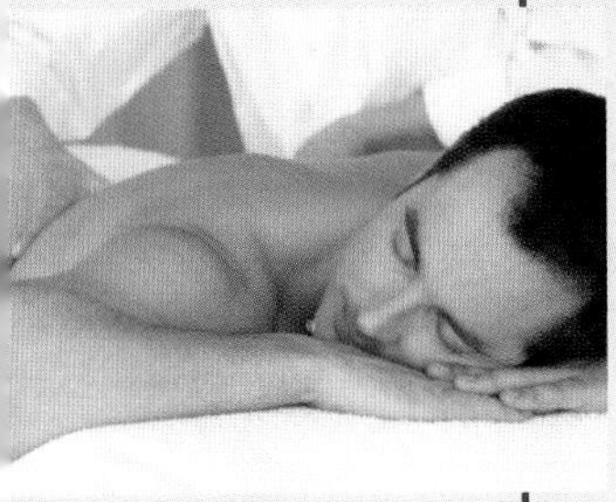

2 Healthy bones and muscles for life

TAKE CHARGE OF BONE AND MUSCLE HEALTH

Your bones and muscles enable you to go about your daily life by supporting your body through the whole range of its activities. Although medical professionals are on hand if there is a problem, you are in the best position to assess if your bones, joints, and muscles are feeling good or overstretched.

41

Knowing how your bones and muscles feel when they are working well and when they are under stress is the first step to preserving their health.

45

People with strong bones and muscles are able to embrace life to the fullest, taking advantage of all the work and social opportunities it offers.

47

Taken appropriately, most medications bring enormous benefit, but some may have side effects on the bones.

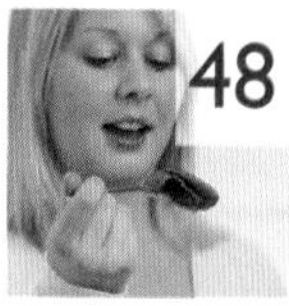
48

There is a lot you can do to treat minor injuries and even some chronic conditions at home, but knowing when to seek help is vital.

51

Being overweight stresses the joints and muscles, but getting the body moving more each day should help to shed a few pounds.

Knowing what's normal for you

From time to time, we all put up with a certain amount of underlying pain in our everyday lives, whether it's caused by tension or aches in the body. How much pain is considered normal? Do you have to grin and bear it all?

FEELING STIFF AND SORE IN THE MORNING

Morning stiffness is not just a sign of aging. Even the young and fit can wake up feeling sore and exhausted. Occasional stiffness can be caused by a disturbed night's sleep after you've been restless, tossing and turning. But if it is a long-term problem, you can take steps to help yourself.

As you sleep, your body repairs the everyday wear and tear on your muscles and bones. In order to do this properly, both your body and mind must be completely relaxed. You can help ease your body into a relaxed state before sleeping by having a warm bath or shower and a hot milky drink. These will relax your muscles, which in turn will help your joints relax.

Although morning stiffness can be a sign of an underlying illness such as arthritis, it is often a warning that the body is being underused. The sedentary lifestyles that many older people lead are an open invitation to morning aches and stiffness. Tendons lose flexibility, range of motion decreases, and muscles shrink. This is why it is important to keep active.

HOW'S YOUR POSTURE?

It can be difficult to distinguish where poor posture ends and spinal disorders begin. A normal upper back is convex, but many people slouch and have rounded shoulders. This is only considered a disorder when the curve becomes excessive. The neck and the lower back are naturally concave, but if this is reversed, it could indicate a more serious condition.

If your spine curves as it should and you have never sustained any spinal injuries but you still suffer back and shoulder pain, the answer may be to improve your posture (see page 72). Think also about how you use your body in your daily tasks. For example, sitting at a computer for long periods of time on a chair that is at the wrong height or the wrong angle will place strain on your joints and muscles, resulting in stiffness and pain (see page 74).

WEAR AND TEAR

Many people around age 50 or over may suffer from a painful, stiff neck with tender muscles, a sensation described as "grating," shoulder and arm pain, and numbness or "pins and needles" in the fingers. A doctor might prescribe drugs and physical therapy to reduce the pain and any inflammation. If symptoms are severe, a collar to support the neck will ease the pressure on the vertebrae until the inflammation subsides. However, much of the wear and tear that causes this sort of pain can be avoided by simple measures, such as good posture and exercise.

Don't dwell on your aches and pains
If you become more active in your daily life and have a positive attitude about what you are doing, you'll be less prone to joint pain and muscle stiffness.

GOOD PAIN, BAD PAIN

Almost everyone feels some pain during exercise, but it is important to distinguish between good pain—the type that's part of the muscle-strengthening process—and bad pain that may indicate an injury.

You may feel some stiffness or soreness for the first 10 minutes of an exercise session while your body warms up. Delayed onset muscle soreness from 4 to 48 hours after exercise is also normal. Although sore muscles are part of the muscle-strengthening process, sore tendons and painful bones are not. These are more likely to be a chronic problem. Use the RICE treatment (see page 95) and see your doctor.

1

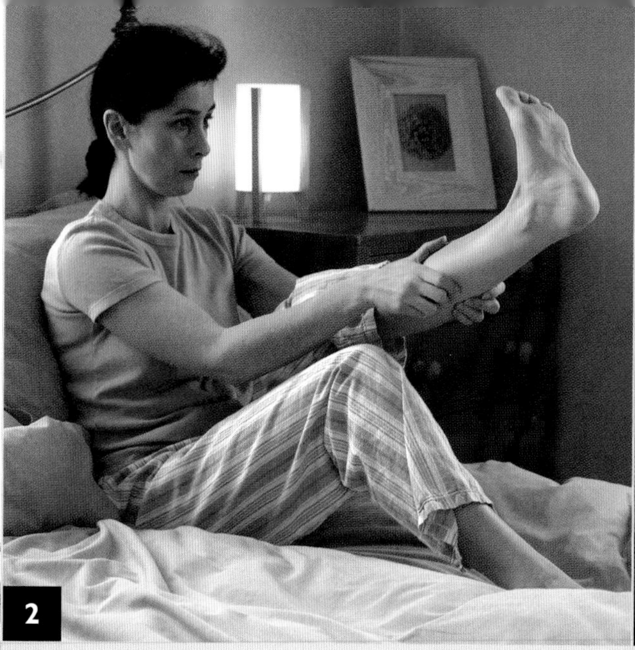
2

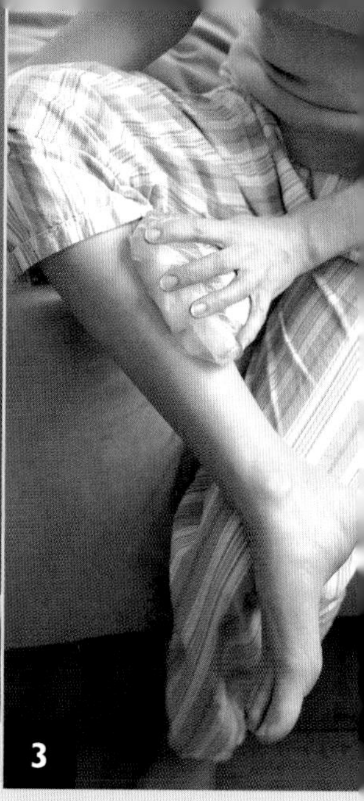
3

Muscle cramps in bed

Some people experience persistent nighttime muscle cramps. If the muscles do spasm, there are a number of things you can do:

1 For calf pain, stand up, put your weight on the pained leg, and bend your knee slightly. For a thigh cramp, keep both legs straight and lean forward at the waist. Use a support to steady yourself.

2 Stretch and massage the area. Straighten your leg and point your toes upward while you gently knead and rub the leg to help the muscle relax. Use circular motions to increase circulation.

3 Apply cold or heat to the leg. Use a warm towel or hot water bottle while the muscle is tense and tight. A cold pack will relieve any residual pain or soreness in the muscles.

EVERYDAY CONDITIONS TO WATCH OUT FOR

Nearly everyone has experienced everyday stresses and strains on their bones and muscles. These often can be relieved by simple techniques.

CRAMPS

These painful muscle contractions can happen to anyone, no matter how fit and healthy he or she is. Cramps commonly occur while you are lying in bed, during and after exercise, or because of prolonged immobility.

Blood is continually pumped into the body's tissues and then propelled away by muscle contractions. If we stand or lie fairly still for a long period of time, the lack of muscle contractions can mean that blood is not being pumped away efficiently, and the circulation slows. This causes an accumulation of carbonic gas and lactic acid in the tissue, and a buildup of glucose, ions, and amino acids. Excess lactic acid can trigger a sharp pain and a strong, sustained contraction of the muscles, causing cramping.

Cramps can be eased by a few simple steps (see above). Drinking a glass of tonic water is also recommended because this contains quinine, which has been shown to be effective in reducing cramping.

Sport-related cramps

Anybody who has watched football or soccer will have seen the players lying on the ground with an assistant elevating their legs and massaging them. The gravity and massage promote blood circulation through the muscles. Soccer players have good hearts and circulation, so their leg cramps result from excessive production of lactic acid and carbonic gas, caused by the intense activity during the game. Stimulating the blood flow to carry away this buildup brings relief from cramps, as does a good rest.

When cramps can be life-threatening

Getting a cramp while you are swimming in cold water can be dangerous. Low temperatures make the blood vessels contract, diminishing blood flow even more. Even on relatively warm days, the body can lose heat fast, because literally all of the body is in contact with the cooler water. A severe painful cramp in this condition can bring a very real risk of drowning.

Growing pains—in the front of the thighs, in the calves, and behind the knees—are a normal occurrence in about 25 percent of children.

GROWING PAINS

During childhood, the body goes through an amazing series of changes. The body changes proportion many times, particularly between the ages of 3 to 5 and 8 to 12. During these growth spurts, children may complain of nighttime leg pains, so-called growing pains.

Growing pains do not hurt in the bones or joints, only in the muscles. This is because the muscles or tendons are still a little too tight for the growing bones. The pain is caused by muscle spasms lasting for up to 15 minutes. Most children don't experience the pains every day; they usually happen once a week or more infrequently.

In most cases, simple daily stretching exercises can help prevent the pain. These exercises must be continued even after the pain subsides in order to keep the muscles and tendons relaxed and able to accommodate the next growth spurt. During a pain episode, stretching the foot and toes upward will often resolve the muscle spasm. Gentle massage and moist heat over the painful spot can also be helpful.

SHIN SPLINTS

"Shin splints" is an imprecise term that covers a range of conditions affecting the shins, such as muscle inflammation, inflammation of the muscle–bone connection, tendinitis, and stress fractures. These conditions result from overuse: doing too much walking or running too soon on too hard a surface. To treat shin splints use the RICE treatment (see page 95). To prevent shin splints,

- Wear well-fitting shoes with heel counters and shock-absorbing soles (see page 84).
- Gradually build up the strength of the muscles in the lower leg so they can withstand the impact of running or long-distance walking. Good exercises include walking on your heels and toe raises.
- Stretch the calf muscle and the Achilles tendon with some slow lunge stretches before exercising.

CRICK IN THE NECK

A crick in the neck is a distinctive combination of pain, stiffness, and a seemingly mechanical limitation of movement—as if something is catching or sticking when you try to move your neck. The pain is mostly caused by protective muscle spasm rather than by an actual mechanical problem. It is usually a neurological overreaction to a relatively minor problem in the soft tissues of the neck—the muscles, ligaments or nerves. You might confuse a crick in the neck with more ordinary pain and stiffness in the neck muscles caused by stress and poor posture, but a crick in the neck is a more distinct problem that comes on suddenly.

Most neck cricks are mild and resolve within a few minutes or a few days at most. However, a malfunction in the joints of the spine can predispose you to chronic pain, further injury, or arthritic degeneration. If you have a crick in the neck that doesn't seem to be going away as quickly as it should, you should see your doctor.

There are things you can do to ease the pain of a crick in the neck:

- Warm your neck using a hot water bottle or hot moist washcloths or in a hot shower.
- Avoid cold drafts. They are one of the most common causes of neck spasm.
- Relax. Stress causes and aggravates cricks, so take time to unwind.
- Keep your neck moving gently. Early neck mobilization significantly improves recovery time.
- Lie on your back and get a friend to gently pull your head away from your shoulders. This should be held for a minute and then slowly relaxed.
- Avoid odd neck positions such as pushing your head forward at the computer or when driving.

Do fish oils really lubricate the joints?

The "old wives" were partially correct here. Oil, specifically fish oil, can benefit inflamed, painful joints, though it doesn't actually lubricate them. Fish oil contains nutrients that help suppress the chemicals in the body responsible for pain and inflammation, and so they relieve joint tenderness, swelling, and stiffness. If you rub fish oil directly onto a painful joint, a certain amount will be absorbed into the bloodstream, but for effective relief, the oil must be taken orally.

ASK THE EXPERT

Dairy milk is a valuable source of calcium during pregnancy for both the mother and the developing baby.

CARPAL TUNNEL SYNDROME

Swelling caused by water retention can be triggered by changing hormonal levels—for example during pregnancy or menopause—and this can affect the bones and muscles, as well. There is a median nerve that, along with flexor tendons, passes through a narrow canal in the front of the wrist known as the carpal tunnel. There is very little room inside the carpal tunnel, and if the tendons are swollen, they can exert pressure on the median nerve, resulting in tingling and a sensation of pins and needles in the area of the hand where the nerve spreads out. This happens particularly at night or during periods of prolonged gripping, of a steering wheel for example. After a time these symptoms will mostly disappear, but sometimes they may persist and require surgical treatment later on. Symptoms can be eased by wearing a wrist splint at night, using a padded wrist support while working at a computer, and avoiding overuse.

THE CHANGES IN PREGNANCY

Several alterations occur within the musculoskeletal system to support the changes in body shape and weight during pregnancy. In addition, different hormone levels in the body during pregnancy affect the bones and muscles.

Shifting center of gravity

Carrying extra weight on the front of the abdomen throws a woman's center of gravity forward, increasing the stress on the lumbar segment of the spinal column—the spine between the lowest ribs and the hipbones. The increased stress is greater than the increase in weight because of the "lever effect." To compensate, the muscles on either side of the vertebrae have to work harder to share the distribution of weight with the increased curvature of the spine. This in turn adds stress to the joints between the vertebrae. All this shifting and weight redistribution can easily lead to lower back pain.

Alterations to the pelvis

The pelvis adjusts its shape during pregnancy because the increasing weight of the baby exerts a force on the front and the joints of the pelvis. The hips are also pushed apart, causing the pelvis to widen. Some discomfort is probably unavoidable, especially toward the end of the pregnancy. However, a fit body with strong muscles and bones will be better able to adapt to these changes.

The need for more calcium

Pregnancy is a time of a high demand for calcium as the baby is developing. To ensure that she does not suffer from a temporary form of osteomalacia (see page 145) at this time, a woman should be sure to eat plenty of calcium-rich foods.

The benefits of strong bones and muscles

Healthy bones, ligaments, and muscles are essential for us to participate in normal daily activities. A flexible body with good muscle tone and strong bones makes us feel energized and enthusiastic about meeting life's challenges.

USE IT OR LOSE IT

The musculoskeletal system supports and protects all the organs of the body. It has to be in good working condition in order for the body tissues to be healthy; the brain and nervous system to be active and responsive; and the circulatory, digestive, and lymphatic systems to be efficient. The ability to move freely is the key to lifelong good health. This will only be possible if your bones, muscles, joints, and ligaments are in good working order.

Too much time spent lounging on the couch will cost you more than you think. Circulation slows down, so vital nutrients like oxygen reach the areas they are needed more slowly, and because waste is not being swept away efficiently, it builds up in the tissues. Without regular use, joints will seize up and will feel stiff when they are called upon to move.

THE BENEFITS OF FREEDOM OF MOVEMENT

Having strong bones and supple muscles means that you will be able to move freely and take hold of what life has to offer. Even an activity as basic as a walk in fresh air will lift your spirits. If you make physical activity part of your life and maintain your ability to move freely, you will feel and look younger and more confident. And if a new challenge that you want to try comes along, you will be able to rise to its demands with confidence. As long as you keep your body in good physical condition, age should not get in your way if you want to take up, say, dancing or sailing at 50 or 60.

STAY ACTIVE—STAY INDEPENDENT

As we get older, the benefits of strong bones and supple muscles are even greater. Later in life, mobility and energy tend to decrease, and the general wear and tear of living can begin to take its toll. An elderly person's joints may feel stiff and weak as his or her bones become less dense, more fragile, and more prone to fracture. But there is no reason to bow to this process as inevitable and unavoidable.

As long as you still have freedom of movement, you will be able to take regular exercise, which will help maintain joint mobility, increase the strength of muscles and bones, and lubricate your ligaments and cartilage. This means that muscles will ache less and will also help ward off degenerative diseases like osteoporosis. Keeping joints healthy helps prevent them from wearing out and thus helps delay or prevent the need for replacement surgery. If you can maintain the ability to move and get exercise in later life you will also help your body make repairs and fight off pain.

MANAGING PERSISTENT CHRONIC PAIN

The benefits of a healthy musculoskeletal system are also important for people suffering from long-term chronic pain. Many people suffering from persistent pain become stressed, anxious, and depressed. It has been shown, however, that exercising stimulates the production of endorphins, natural chemicals that can raise tolerance of pain, lift mood, and relax the muscles. Not only is the pain reduced, but the sufferer can also achieve a more relaxed state of mind and body, which will help put pain into perspective and strengthen the ability to cope.

In the swing
Being able to stay active all through life will ward off the effects of age and long-term use on the bones and joints.

ENJOYING FREEDOM OF MOVEMENT

As you move—during a walk, for example—your body temperature rises by three to five degrees, giving you a feeling of warm well-being. Stress chemicals that accumulate in the tissues during inactivity burn off, and your body's natural feel-good hormones, the endorphins, are stimulated and released into your system. After a good walk, your body's natural instinct to regenerate itself takes over, putting you into a state of relaxation.

Being able to move freely and getting regular exercise have psychological as well as physical advantages. The physiological changes in the body and brain can induce a sense of elation when we exercise and anxiety and depression when we do not.

If you are very unfit, it may take several weeks of exercise before you see and feel a positive change in your body condition. But no matter how out of shape you are, exercise will lift your mood instantly. Self-esteem, too, has been shown to improve after just one session.

KEEPING JOINTS HEALTHY

Protecting your joints is an effective way to avoid or relieve pain and to prevent further damage to your joints. You can minimize joint problems by following a few good lifestyle habits.

Use good body mechanics

- Stand up straight. Good posture protects the joints in your neck, back, hips, and knees.
- When you are standing, the height of your work surface should be adjusted to minimize stooping.
- When you are sitting, the correct height for your work surface is determined by measuring your elbow at 90 degrees.
- When you pick items off the floor, bend your knees and hips. If you can, sit on a chair and bend over.

For more details on the correct way to carry your body when standing, sitting, or lifting see pages 72–75.

Use your hands wisely

- Avoid twisting or using your joints forcefully. Instead of straining to open a jar, use hot water to loosen the lid and your palm to open it or use a jar opener.
- Avoid positions that push your fingers toward the little finger.
- Try holding objects with an open hand rather than gripping them.
- When opening drawers, using a loop that you pull with your wrist or forearm can decrease stress on your fingers.

Keep your joints moving

- Once daily, move each joint through its full range of motion, without moving so far that you feel any pain. Move slowly and gently—sudden jerking can hurt the joints.
- When writing or doing other work with your hands, release your grip every 10 to 15 minutes.
- On long car trips, get out of the car and stretch at least every hour.
- On a long flight, move around the cabin occasionally to avoid stiffening joints.
- When traveling on a train, walk from car to car to stretch your legs.

Avoid overworking your joints

- Avoid rushing. Work at a steady, moderate pace.
- Rest before you become fatigued or sore.
- Alternate light and moderate activities throughout the day.

Listen to your pain

Learn to distinguish between normal everyday aches and pain from the overuse of a joint. If pain lasts for more than an hour after an activity, it may have been too stressful. Next time, try to do a little less or find a way to do it that takes less effort.

Medication and the bones

Many people are aware that being female, having a low body weight, and leading a sedentary lifestyle are risk factors for osteoporosis. However, osteoporosis can also result from the long-term use of certain medications.

Some prescription medications, when used on a long-term basis, can decrease bone mass and bone strength. Often, drugs are the only way to manage a serious or chronic health condition. In many cases, they are life-saving or vital for maintaining quality of life. But in the process of treating one condition, some drugs may cause another: osteoporosis. Osteoporosis resulting from medication is known as secondary osteoporosis.

If there is no alternative drug available, it is recommended that the lowest dose of medication is prescribed for the shortest period of time. It is important to discuss the risk of osteoporosis with your doctor: Do not stop taking or alter your medication dose on your own.

CORTICOSTEROIDS AND BONE LOSS

Corticosteroids are antiinflammatory drugs used to treat a variety of inflammatory and allergic conditions, such as rheumatoid arthritis, asthma, chronic liver disease, and inflammatory bowel disease. Although they are effective in treating these conditions, with long-term use they are the most common cause of drug-induced osteoporosis.

Corticosteroids decrease the amount of calcium absorbed from food, increase calcium loss from the kidneys, and decrease bone formation. They also interfere with the production of sex hormones in both men and women, which can contribute to bone loss and cause muscle weakness. The bones most at risk from osteoporosis are in the spine, the hips, and the wrist.

Patterns of bone loss

Corticosteroids are administered in a number of different ways: orally in tablets or pills, by injection into the joints, or by inhaler. Bone loss tends to be greatest in people using oral forms at higher doses and over longer periods of time. Inhaled corticosteroids and other local forms of these preparations are less likely to cause bone loss, so doctors may prescribe these first.

REDUCING THE RISKS

Corticosteroid-induced osteoporosis is both preventable and treatable. Because bone loss occurs most rapidly in the first 6 to 12 months of therapy, prevention measures should begin early, ideally at the onset of drug treatment. Bone health must be carefully monitored, so individuals starting corticosteroid therapy should have a bone mineral density test (see page 108). This will provide a base measurement from which to monitor subsequent changes in bone mass.

Lifestyle plays an important role in the prevention and treatment of drug-induced osteoporosis. Ensure that you are getting adequate levels of calcium and vitamin D, and make lifestyle changes, such as not smoking and drinking alcohol only in moderation, to ensure you are not stripping the body of bone.

Other drugs that can cause bone loss

In addition to corticosteroids, there are a number of other drug types that can cause bone loss. The drugs listed below come in many forms and are known by a variety of generic and brand names, so be sure to ask your doctor whether your medication falls into one of these categories.

Aluminium-containing antacids
Antacids neutralize stomach acid and are widely used for the treatment of indigestion. Aluminium antacids should not be taken at high doses for long periods of time because they can cause weakness and bone loss.

Anticonvulsant (antiseizure) drugs
Some anticonvulsants lower the calcium concentration of the blood and subsequently disturb bone metabolism. This can lead to rickets and osteomalacia (see page 145), with accompanying pain and weakness.

Thyroid hormones
Long-term thyroid-hormone replacement therapy for hypothyroidism can cause bone removal to exceed bone formation.

Gonadotrophin releasing hormone (GnRG) analogs
GnRG analogs decrease estrogen production and are used to treat endometriosis. Because of the estrogen loss, these drugs increase bone loss, although this stops once medication is stopped.

Heparin
Heparin is used to prevent blood clotting. It has been reported that prolonged use can increase the rate of bone breakdown and impair bone formation.

Self-help for bone and muscle problems

Although they may be painful, many conditions affecting the bones and muscles are considered minor, self-limiting, and easy to treat at home. Even more serious conditions—after advice from a doctor—also respond to self-help measures.

Bone and muscle problems range from the occasional twinge after a strenuous afternoon of gardening or the odd football game, to the debilitating conditions of osteo- and rheumatoid arthritis.

SPORTS INJURIES AND EVERYDAY ACCIDENTS

In sports activities or after a mishap such as a fall, injuries to joints, muscles, bones, tendons, and ligaments are common.

- Overstretched or torn fibers in a muscle or tendon (which attaches muscle to bone) are called strains. Catching your thigh on a piece of furniture, for example, tears a few muscle fibers and causes bruising.
- A sprain occurs when a ligament becomes overstretched and ruptures. Sprains account for most injuries to the ankles.
- Cartilage, particularly in the knee, can be torn by a twisting fall, for example when skiing, playing football, or running.

Accident prevention

It pays to avoid injuries in the first place.

- Get familiar with equipment: Bad usage is a main cause of accidents. When taking up a new sport, seek coaching on good-quality equipment, including safety issues.
- Always wear appropriate clothing, and follow advice on safety gear, such as pads, guards, and helmets.
- Always warm up properly before taking part in any sports activity. It is equally important to perform some cool-down stretches after exercise.
- For sports in general and for contact sports in particular, physical fitness is an important preventive factor—the fitter you are, the less likely you are to sustain an injury.
- Be aware that tired muscles injure more readily than fresh ones.

How you may feel

Strains, sprains, and damage to cartilage cause bleeding, bruising, swelling, and pain. Their severity will depend on the degree of the damage. Pain is caused by damage to the nerve fibers. This can last from five to seven days, although it should ease a little every day. The nerve endings are also irritated by toxic substances expelled from the injured tissue and the pressure caused by the increased flow of blood to the area. This increased blood flow clears away the poisonous waste substances but causes swelling. Redness is caused by the dilation of small blood vessels as they widen to increase the supply of nutrients and healing substances.

What you can do

If there is any suspicion of a broken bone, a trip to hospital is advisable so that X-rays can confirm this or rule it out. For minor injuries, follow the RICE procedure to relieve pain, minimize bleeding, and restore

Time to relax
Being in pain can be mentally as well as physically debilitating. Learning to relax, perhaps in a warm scented bath, can make it easier to handle pain.

Fresh pineapple is frequently recommended by nutritional therapists to help heal sports injuries. It contains the enzyme bromelain, which is thought to accelerate tissue repair.

function to the joint (see page 95). Ice will reduce bruising for the first couple of days, after which heat will speed healing. An injured joint or limb can be supported with a firm, but not overtight, bandage.

More severe injuries need rest, and a doctor may also recommend painkillers, antiinflammatory drugs, and physical therapy. You can begin to move the affected part gently once the pain has subsided.

MANAGING OSTEOARTHRITIS

The most common form of joint disease is osteoarthritis. Men are more likely to develop osteoarthritis before the age of 50, but after the age 50, it is more prevalent in women. Osteoarthritis sets in when the cartilage that protects the ends of bones at their point of contact degenerates, causing the bare bone to thicken and widen.

The bone underneath the cartilage is well supplied with nerve fibers, so as the area is exposed it becomes very painful. The movements of the joints become limited, mainly because of the loss of lubrication and the grinding between the two exposed bone surfaces. In addition, bony outgrowths—bone spurs or osteophytes—may sometimes form.

The degeneration may set in as a result of aging, wear and tear, overuse, or injury to the joint. It may also occur after rheumatoid arthritis or gout. Obesity may be a factor, along with repetitive use of the joint.

How you may feel

Ostearthritis may cause mild pain, swelling, stiffness, and a "creakiness" in the joint, or it can result in a total loss of function with muscle wasting, ruptured tendons, and excruciating pain. It tends to affect the large weight-bearing joints, such as the hips and knees. However, the small joints in the hands, feet, and fingers, as well as the vertebrae in the neck and lower back, can also be affected.

What you can do

Although there is a lot you can do to help yourself, see a doctor first: Osteoarthritis is not something to self-treat without supervision. Drugs and other treatments may slow down and even interrupt the condition's progress, but they cannot repair the lost cartilage or slow down the degeneration. In the earlier stages, painkillers, antiinflammatory drugs, heat treatment, and physical therapy can help relieve pain and maintain mobility. If the condition is advanced, steroid injections can offer some ease.

Losing weight will relieve pressure on the joints. Gentle, non-weight-bearing movement will help keep a joint mobile for as long as possible by stimulating the circulation of synovial fluid while keeping the ligaments and muscles supple. This helps preserve range of movement.

Avoid activities such as lifting or walking over rough terrain, which put extra pressure on the weight-bearing joints. Swimming may be beneficial because it strengthens the muscles supporting the joints without putting any pressure on them. Bicycling is another good option. Walking with a stick will ease the pressure on affected weight-bearing joints as well. The stick should be used in the hand opposite the affected side, so that the body weight is

IT'S NOT TRUE!

"Supplements from exotic species will help"

Extracts of shark's fin (obtained from the food-processing industry), rhinoceros horn, and other exotic concoctions, often promoted as beneficial for osteoarthritis, are extremely unlikely to be of any help. Not only does their trade pose a threat to a number of species, but the preparations themselves are likely to contain dozens of substances in low and ineffective doses. The body will simply break down and digest molecules like proteins, such as those found in fish, into amino acids and absorb them as food.

Self-help for arthritis

Diet *Eat healthily and make an effort to lose any excess weight to relieve stress on your joints.*
Exercise *Work out an exercise routine for yourself to make the most of your range of movement.*
Sleep *Get plenty of quality sleep to counter fatigue induced by arthritis.*
Relaxation *Learn relaxation techniques to help beat stress.*
Pain management *Your doctor can give advice on the latest medication to relieve arthritis pain and how to use it effectively.*
Communication *Try not to let arthritis take over your life: Keep up social contacts by phone or email if pain keeps you at home.*
Mental well-being *Learn to recognize the signs of depression and get timely help to treat it. Training in stress management may also be beneficial to some sufferers.*

divided between the stick and the healthy leg. A nurse or physical therapist can help you get the synchronization right.

Chronic pain is mentally, as well as physically, debilitating. Practice relaxing; this has been shown to help counteract the muscle tension that accompanies stiff and painful joints. Try devoting time to pleasant activities that you really enjoy, such as listening to music, yoga, or meditation, and if the condition allows, gardening, gentle walking, and swimming.

COPING WITH RHEUMATOID ARTHRITIS

Rheumatoid arthritis is a chronic disease of the musculoskeletal system that affects the joints in the hands, wrists, knees, and feet. It generally strikes people between the ages 40 and 50, with women having a three times greater risk than men. In this condition, the membrane lining a joint capsule becomes painfully inflamed and swollen making the joint stiff, particularly after a night's sleep. The sufferer can also experience muscle weakness and fatigue. Rheumatoid arthritis is an autoimmune disease, in which the immune system attacks the body's own tissues. The exact cause of this is unknown, but viral infections, stress, diet, and bacterial overgrowth in the digestive tract are thought to be contributing factors.

How you may feel

Early symptoms may include a vague aching in the muscles and generally feeling unwell. As the condition progresses, joint pain and swelling occur. The knuckles, toes, and other small joints are most susceptible, but larger joints may also be affected. In some cases, the onset is sudden. A mild form of anemia usually accompanies the condition. In severe cases, the inflammation spreads to the nearby cartilage and bone, causing deformities and disability.

What you can do

A physical examination and blood tests are needed to establish the presence of rheumatoid factor. X-rays will reveal the extent of the damage. Treatment may include physical therapy, heat treatment, painkillers and antiinflammatory drugs. Joint replacement surgery may be used if the damage is severe.

Gentle movement and manipulation can assist mobility in the affected joints.

ON THE CUTTING EDGE

Vitamin C and antioxidants

According to the American Arthritis Foundation, simply increasing consumption of oranges and other citrus fruits or the juice of these fruits may be of benefit to arthritis sufferers. Recent research has shown the importance of vitamin C and other antioxidants in reducing the risk of osteoarthritis and its progression. Citrus fruits (oranges, lemons, limes) are also good sources of folic acid, which can help alleviate the side effects of rheumatoid arthritis drugs such as methotrexate and reduce the risk of cardiovascular disease in women who have lupus.

Why weight matters

Weight loss decreases stress on the bones and joints, and this can improve the symptoms of osteoarthritis. Weight loss also provides extra energy and flexibility, which help build up muscle strength.

Being overweight stresses the body and results in increased musculoskeletal problems, such as

- lower back pain;
- hip and knee problems;
- osteoarthritis;
- disc problems; and
- fractures or sprains of feet or ankles.

LOWER BACK PAIN

Excess weight puts stress on the lower back, the joints, and the discs. Most excess weight is found around the abdomen. This puts tremendous pressure on the lower back as it is forced to curve forward to adapt to the weight it has to bear. As the weight of the upper body falls on the base of the spine, these bones are at increased risk of wear. Thus, weight gain around the abdomen ultimately damages the structures of the lumbar spine.

HIPS, KNEES, ANKLES, AND FEET

Most of our body weight is borne by the hips, knees, ankles, and feet, which are affected by wear and tear more frequently in obese people, leading to arthritis. Other problems include:

- Muscle imbalances, particularly in the muscles that support the foot and ankle.
- Increased alignment problems for lower extremities—foot arch collapse, knock knees, and excessive knee and hip joint rotation.
- Reduced shock absorption, leading to cartilage damage.

In order to assist the joints, people who are overweight should wear laced shoes with a strong heel counter (see page 84) to reduce the heel strike shock caused by the excess weight on the joint cartilage.

OSTEOARTHRITIS

Cartilage acts as a cushion or shock absorber between the bones. When cartilage breaks down in osteoarthritis, the protective cushion is lost and the bones grind together. This causes the symptoms of pain, swelling, and restricted motion. Most of the weight of the body is carried by the hips, knees, ankles, and feet. Being overweight increases the risk of osteoarthritis by placing extra pressure on these joints and wearing away the cartilage.

Ten extra pounds of weight can increase the force on the knees by 30 to 60 pounds with each step.

It has been calculated that the risk of developing osteoarthritis may increase by 9 to 13 percent for every 2-pound increase in weight. In other words, being 20 pounds overweight can double your chance of getting osteoarthrtis. Losing weight can help prevent osteoarthritis or relieve symptoms and slow the progession of the

Keep on moving

Keeping to a healthy weight does not have to involve hours in a gym. Most of us could make more of the exercise opportunities in our daily lives: walking rather than driving and using stairs rather than elevators or escalators.

INCREASING ACTIVITY FOR WEIGHT LOSS

The best way to lose weight is through a combination of a healthy diet and exercise (see page 86). Another excellent contribution to weight loss is to build extra activity into your daily routine.

Extra effort
When you wash your car or clean your windows, use large, circular movements to increase your range of motion.

Expansive movement
Exaggerate your movements when you are drying yourself after a shower or getting dressed.

Move while relaxing
Do some exercises while watching TV. Stretch during the commercials, fold laundry, or make a drink. Forget the remote control—get up to change channels.

Walking and stair climbing
Walk longer distances: Get off the bus a stop early. Walk up and down stairs instead of taking the elevator.

disease. It is never too late to start losing weight. If you already have arthritis, losing just a few pounds can significantly reduce pain and in some cases can eliminate the need for surgery.

RESULTS OF SURGERY

Joint replacement surgery tends to be less successful in obese people to the extent that some surgeons refuse to perform such surgery on people who are severely overweight.

Patients should make every effort to reach normal weight before surgery. This should lessen postoperative complications and prolong the life of the new joint. If a patient puts too much stress on the new joint, it can cause fracturing of the prosthesis or loosening of the prosthesis from the bone.

OBESITY AND MUSCLE WEAKNESS

There is a strong relationship between obesity and muscle problems. Muscle support and strength is directly weakened by excess weight, and this can lead to poor posture and decreased flexibility. This can become a "catch-22" situation. Obesity can lead to increased difficulty with leisure activities, housework, and exercise (walking, climbing stairs, and standing up). Increased tiredness and fatigue while performing these activities may lead to a decreased activity level, which can lead to an increase in weight, which will lead to further muscle weakness.

THE IMPORTANCE OF INCREASED ACTIVITY LEVELS

It is essential to keep active and to maintain muscle strength, because muscular weakness can have a huge impact on the function and condition of the musculoskeletal system. It can contribute to lack of balance and coordination, falls, and fractures. Even a short period of immobility can have an impact on muscle health. Extended periods of immobility can result in joint tightening, muscle contracture, and atrophy.

After only a week of bed rest, muscles can lose as much as 10 to 15 percent of their strength.

Losing excess weight can help maintain energy levels, flexibility, and muscle strength. Increased strength and endurance leads to increased independence, which in turn leads to increased involvement in activities and improved quality of life.

EXERCISE

One problem is that osteoarthritis often makes it painful to exercise, leading many sufferers to avoid physical activity and put on more weight. Low-impact exercises, such as those on page 93, can help sore joints and relieve stiffness and swelling. An exercise plan that is safe, gradual, and tailored to the individual's needs and interests will bring real benefits. The first step is to speak to your doctor and then devise an exercise plan under the supervision of a therapist or qualified trainer. See pages 92–94 for information on exercising with osteoarthritis.

When diet and exercise are combined for weight loss, they bring about other changes, including better stress management, improved energy levels, and a decrease in the risk of developing serious health problems.

FOOD AND DRINK FOR STRONG BONES

There's a lot you can do to improve and preserve bone health. Knowing the most important nutrients and the best dietary sources is a good first step.

Feeding your bones

The health of your bones depends on a range of physical and lifestyle factors—not least of which is the inclusion of essential nutrients in your diet. These nutrients influence the development and maintenance of strong and healthy bones.

Bone is living tissue, which means that it is continually being formed and broken down (resorbed). The adult human body contains between 2.25 and 3.5 pounds of calcium, most of which is in the skeleton. Calcium, and other substances from the diet, are deposited in the bones and teeth. The mineral content and density of the bones increase from infancy and peak in early adulthood.

During the first two decades of our lives, bone grows both in length and width. The formation of bone dominates over bone resorption, and bone mass steadily accumulates. About half of our peak bone mass is reached during the adolescent growth spurt and, at the end of puberty, bone continues to become thicker and denser. Peak bone mass is usually reached in the 20s, when the bone is at its strongest.

After this point, bone loss begins. Calcium and other substances are removed from the bone more quickly than they are added, and this continues until the end of our lives. Achieving good peak bone mass in early adulthood is important in reducing the risk of osteoporosis later in life, because it means that bones are as strong as they can be before loss begins.

THE IMPORTANCE OF CALCIUM

Calcium is a mineral needed for the contraction of muscles, the healthy functioning of nerves, and the clotting of blood, and to provide the structural support of the skeleton. It also helps the work of several enzymes in the body. Many studies have shown that getting enough calcium reduces the risk of developing osteoporosis, hypertension, and colon cancer.

Where is calcium found?

Calcium is obtained from foods that are rich in the mineral, such as dairy foods, tofu, and some green leafy vegetables. It is also added to some processed foods and drinks.

BIOAVAILABILITY AND ABSORPTION OF CALCIUM

Up to 60 percent of the calcium we eat is absorbed by the body. The amount absorbed depends on a person's health status, how much calcium is needed, the amount of calcium in the diet, and the presence of compounds, such as vitamin D, that may enhance the absorption of calcium.

In general, the calcium in milk and other dairy foods is much more readily absorbed than that derived from plant sources. Studies have shown that for most foods, between a quarter and a third of calcium is absorbed by the body.

There are several substances that can interfere with the body's absorption of calcium. Some of the most common offenders are foods high in oxalates, phytates, protein, and sodium. Foods high in oxalates include spinach, rhubarb, chocolate, baked beans, and peanuts. Legumes such as pinto beans and peas are high in phytates. The calcium in legumes is only half as available as the calcium in milk. Wheat bran is also high in phytates and is the only fiber-rich food that appears to reduce calcium absorption—the fiber in

Dairy is best

Dairy foods are the richest source of calcium in the Western diet. One 8-ounce glass of milk provides as much calcium as the following:

- 12 slices of whole-wheat or white bread
- 6 oranges
- 6 canned sardines
- 1½ cups of blackeyed peas
- 3 tablespoons of sesame seeds
- 2 slices of cheese pizza
- 4 slices of french toast

fruits, vegetables, and common cereals does not interfere with calcium absorption. To derive the maximum benefit from your calcium-rich foods, do not eat them at the same time as foods high in oxalates and phytates. If you wish to eat foods from these categories, eat them one hour before or two hours after calcium-rich foods.

Excessive protein and sodium intake can increase calcium loss through the kidneys. In fact, an individual's daily calcium requirement increases in direct proportion to the amount of protein and sodium in the diet.

FEMALE TEENAGERS

Pregnant teenagers need more calcium

Most expectant mothers do not need to increase their calcium intake during pregnancy, particularly if they are of average weight and eat a balanced diet. Teenage mothers, however, need to consume more calcium than normal for several reasons. First, teenagers are still growing and their bones are still developing, making adequate intake vital. Second, estimates suggest that three quarters of all 16-to-18-year-old girls—pregnant or not—do not eat enough calcium-containing foods.

Teenage mothers should maintain a regular, high intake of calcium, roughly in excess of 1200 milligrams per day. That's at least half a quart of milk, two slices of bread, and a chunk of cheddar cheese; or a can of sardines, a large serving of macaroni and cheese, and an orange.

NON-DAIRY SOURCES OF CALCIUM

Vegans may be more vulnerable to low intakes of calcium, as they do not eat dairy products. However, many soy products, such as tofu and soy drinks, are fortified with calcium and so can be a useful source for vegans. Although the calcium in vegetables is bound by fibers and phytates or phytic acid, which may interfere with its absorption, in practice this may not be a real cause for concern, because vegetables also have a low phosphate content, which tends to improve calcium absorption. Furthermore, most people naturally produce the enzyme phytase, which can break down the phytates and phytic acid in vegetables.

Tofu, or bean curd, is an excellent source of calcium. It is made by solidifying soy milk with the use of the mineral calcium sulphate.

Green leafy vegetables, seeds, and nuts also provide important amounts, and some soy milks are fortified with calcium. In areas of the country with hard water, this can provide about 200 milligrams of calcium a day, but soft water does not contain any calcium. Black molasses, edible seaweeds, watercress, parsley, and dried figs are also good sources.

EATING WELL DURING PREGNANCY

A baby's teeth and bones start to develop from as early as five weeks after conception and continue to grow throughout infancy. An expectant mother needs to ensure that she provides the nutrients necessary to form strong and healthy bones. Just as children, teenagers, and adults need a diet rich in bone-forming nutrients (see pages 58–59), so too does a baby as it develops in the womb. This is one of the significant periods of skeletal growth, during which the bones increase in size and bone mass begins to accumulate. It is during this time, therefore, that the baby must get all the nutrients it needs.

If a woman was of average weight prior to conception and enjoyed a healthy, balanced diet, there are not many changes that she would have to make to aid the healthy development of her child. The absorption of calcium increases—as does that of iron—during pregnancy, and less is excreted in the mother's urine, making it more possible for supplies to be conserved. Although the baby may draw calcium from the mother's bone mass, the baby's calcium needs are unlikely to be at risk, and it should always receive the calcium it needs. Although women do not necessarily have to increase the quantities of calcium-rich foods in their diet, some experts recommend

Studies have shown that most teenagers of both sexes, as well as half of women and a quarter of men, do not eat enough calcium-containing foods on a daily basis.

CALCIUM—ARE YOU GETTING ENOUGH?

Calcium intake requirements vary by age. The following are requirements from the National Academy of Sciences Institute of Medicine, Food, and Nutrition Board. Most food labels do not give calcium content in milligrams, but rather in % DV (daily value). Children 9 to 18 years old should get 130% of the DV.

BABIES

Newborns and babies up to 6 months old require 210 milligrams; those 6 to 12 months old need 270 milligrams. Studies have shown that the calcium in breast milk is more easily absorbed than that in formula, although both provide adequate amounts.

PRESCHOOL CHILDREN

School children 1 to 3 years old require 500 milligrams of calcium daily. According to USDA statistics, up to 30 percent of these children (and those up to 5 years) do not get enough calcium. It is important to establish good eating habits for children at an early age.

ADOLESCENTS

Children 9 to 18 years old need 1300 milligrams of calcium daily in order to optimize their peak bone mass. Studies have shown that most children this age do not get adequate calcium in their diet: Less than half of girls ages 9 to 12 get enough, and this inadequacy continues for most women: Government statistics show that up to 90 percent of girls (and 70 percent of boys) ages 12 to 19 do not get enough calcium. The popularity of soft drinks among children in this age bracket is an additional problem: This type of beverage, which often contains phosphoric acid, can impair calcium absorption and has been linked to obesity.

SCHOOL CHILDREN

Children 4 to 8 years old need 800 milligrams of calcium every day. According to government statistics, inadequate calcium in the diet affects 7 of 10 girls and 6 of 10 boys in this age bracket (and up to age 11).

ADULTS

Adults 19 to 50 years old should have at least 1000 milligrams of calcium daily. This increases for pregnant and breastfeeding women, who should have at least 1200 milligrams and up to 1500. More than 2500 milligrams per day can be harmful and is not recommended for anyone. Most adults—70 percent of men and 90 percent of women—do not get enough calcium on a daily basis.

OLDER ADULTS

Adults over age 50 need 1200 milligrams of calcium daily. Women not on HRT need an additional 500 milligrams. A nationwide survey of adults over 60 found that 70 to 87 percent of them don't get enough calcium, even with supplements.

that expectant mothers increase their calcium intake by as much as twice the pre-pregnant intake, in order to protect the mother's bones from the risk of osteoporosis during this period. Breastfeeding mothers continue to require at least 1200 milligrams of calcium a day after their babies are born. During pregnancy, the mother's body will try to store calcium to produce breast milk. The period of breastfeeding is a time of high calcium demand.

In addition to regular calcium-rich foods, pregnant and breastfeeding women should make sure their diet includes a rich variety of fruit and vegetables, grains, and proteins, to ensure a sufficient supply of all the other basic nutrients required during this time—specifically vitamin D to aid absorption of calcium (see page 58), iron, zinc, and folate, for healthy fetal development.

Vegetarians who eat dairy products and eggs should be getting adequate amounts of calcium and other nutrients from their diet; vegans should consult a doctor or dietitian on their nutrient intake and consider supplements if recommended.

Get the right balance

Gaining too much weight during pregnancy can lead to complications—high blood pressure, prolonged labor, difficulty shedding the pounds after the birth, and varicose veins—but not gaining enough weight, through malnutrition, can have serious lasting implications for the child. A low-calorie intake, coupled with a lack of the right nutrients, can cause pregnancy complications and produce a baby of low birth weight or one that is premature. Compared to normal babies, low birth weight babies

- are at greater risk of death during infancy;
- are more likely to have asthma, respiratory tract infections, and ear infections;
- have increased risk of cerebral palsy (this is especially true of babies weighing less than 2 pounds); and
- are more likely to reach developmental milestones later than average.

A healthy weight gain in a pregnant woman also builds up the fat and fluids she will need when breastfeeding. Current guidelines suggest that expectant mothers should gain 25 to 30 pounds during pregnancy: Those who are underweight at conception should gain slightly more, those who are overweight slightly less. Women who suffer from morning sickness may find that it helps to eat more frequent, smaller meals and snacks—up to six times a day.

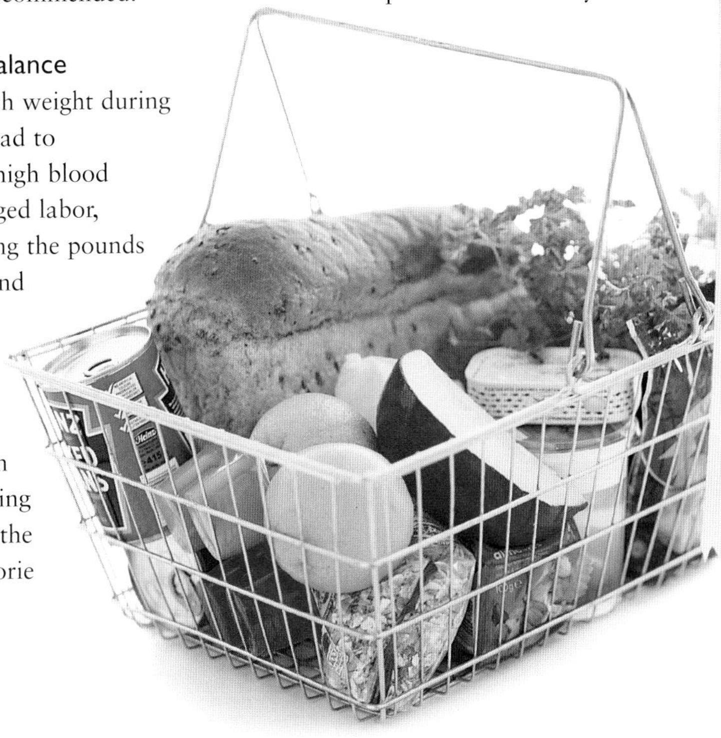

FOODS RICH IN CALCIUM

AVERAGE PORTION	CALCIUM (MG)
DAIRY PRODUCTS	
Macaroni and cheese $1^1/_3$ cups	510
Emmental 1.5 oz.	388
Gruyere 1.5 oz.	380
Cheddar, reduced fat 1.5 oz.	336
Edam 1.5 oz.	308
Whole milk yogurt, plain $^2/_3$ cup	300
Cheddar 1.5 oz.	288
Low-fat yogurt, plain $^2/_3$ cup	285
Cheddar, vegetarian 1.5 oz.	276
Skim milk 6.5 oz.	249
Low-fat milk 6.5 oz.	248
Whole milk 6.5 oz.	237
Rice pudding, canned 7 oz.	186
Cheese spread, plain 2 tbsp.	126
Cheesecake 4.2 oz.	78
BREADS AND CEREALS	
Scones, plain 1.7 oz.	86
White bread 2 slices	79
Brown bread 2 slices	72
Muesli, Swiss-style $^1/_4$ cup	55
Whole-grain bread 2 slices	39
FRUIT AND VEGETABLES	
Kale, steamed $^1/_2$ cup	180
Figs, dried 3	150
Spinach, steamed $^1/_3$ cup	144
Oranges 1 medium	75
Apricots, dried 2 oz.	52
Yam, baked 1 medium	26
Parsley, fresh 1 tbsp.	6
OTHER SOURCES	
Sardines in tomato sauce 3.5 oz.	430
Seaweed, kombu, raw dried .35 oz.	90
Seaweed, wakame, raw dried .35 oz.	66
Baked beans $^2/_3$ cup	65
Soy flour 1 tbsp.	63
Seaweed, nori, raw dried .35 oz.	43
Molasses 1 heaped tbsp.	36
Brazil nuts 6	34
Almonds, toasted 6	31
Soy milk, plain 6.5 oz.	26

Other nutrients for bone and muscle health

In addition to calcium, other important nutrients that are used by the body to maintain the health of your bones and muscles include vitamin D, phosphorus, and magnesium. Some nutrients work together for maximum benefit.

VITAMIN D

Vitamin D is needed for the absorption of dietary calcium from digestion. It also has a direct effect on the deposition of calcium in bone. This vitamin is made via the action of ultraviolet (UV) sunlight on the skin. Depending on time of year and latitude, some people can get enough vitamin D through sun exposure. There isn't a daily recommended amount for adults, but there is an Adequate Intake (AI) level, the level sufficient to maintain healthy blood levels of an active form of vitamin D. For adults 19–50, the AI is 5 micrograms; for adults 51–69, it is 10 micrograms; and for those over 70, it is 15 micrograms.

Some groups of people, such as infants and children, pregnant and breastfeeding women, the elderly, the housebound, and people who stay covered up for most of the year, may not be able to obtain adequate amounts of UV sunlight and so may need dietary vitamin D supplements. Infants and children who are vitamin-D deficient can develop rickets—a condition in which the bones become deformed and are too weak to support any weight. Because this can become permanent, supplements containing vitamin D

Eating a variety of foods

When choosing foods to build and maintain healthy bones and muscles, select foods that are tasty and varied. A well-balanced diet, comprising all the major food groups, will help ensure a wide range of valuable nutrients every day without the need for supplements. It is important to avoid salty, processed foods and foods that are high in sugar. If possible, try to limit caffeine consumption (tea, coffee, and cola) because it can inhibit the absorption of valuable nutrients.

Boron and vitamin K

Boron is required to increase bone density and prevent bone loss, and vitamin K (potassium) is needed for synthesis of proteins found in bone. Oats, raisins, apples, and almonds are good sources of both nutrients.

Potassium

Increased potassium intakes have been associated with higher bone density. Foods high in potassium include vegetables, legumes, fruits, and whole grains.

Vitamin C

Vitamin C is used to make collagen, a protein found in bones, cartilage, and joints. Sources include honeydew melon, oranges, grapefruit, tangerines, and strawberries.

are available to children. Some adolescents, elderly people, and women who are repeatedly pregnant and breastfeeding are at risk of vitamin D deficiency. Vitamin D deficiency in adults can result in softening of the bones, a condition known as osteomalacia.

Few foods contain large quantities of vitamin D. However, fish liver oils such as cod liver oil have very high levels, and oily fish like sardines, salmon, tuna, mackerel, and trout are also rich sources; these foods provide about one third of our dietary vitamin D intake. Meat, poultry, egg yolk, liver, and butter are also important sources. Milk is fortified with vitamin D, as are many reduced-fat spreads, some breakfast cereals, and soy products.

Studies have shown that vitamin D may improve bone density and reduce rates of fracture in the elderly, particularly when taken with calcium. This has been demonstrated among people in residential care.

PHOSPHORUS

Phosphorus is another essential mineral for healthy and strong bones. When combined with calcium, phosphorus—in the form of calcium phosphate—gives strength to our teeth and bones.

The main sources of phosphorus in the American diet are milk and dairy products, grains, vegetables, and meat. As with calcium, milk and dairy products are the most concentrated sources of phosphorus, although meat, fish, nuts, fruits, cereals, and vegetables all provide useful sources. Phosphorus is added to a number of processed foods such as baked products and carbonated drinks. Deficiency of phosphorus is rare, although diuretic drugs in elderly people and overuse of antacids containing magnesium and aluminium may cause deficiencies.

DIETARY SOURCES OF PHOSPHORUS

AVERAGE PORTION	PHOSPHORUS (MG)
Sardines in tomato sauce 3.5 oz.	420
Braising steak, cooked 5 oz.	308
Cheddar, reduced-fat 1.5 oz.	248
Cod, poached 4.2 oz.	216
Chicken, dark meat 3.5 oz.	200
Cheddar 1.5 oz.	196
Low-fat milk 6.5 oz.	190
Skim milk 6.5 oz.	188
Whole milk 6.5 oz.	184
Peanuts, dry roasted 1.5 oz.	168
Chicken, light meat 3.5 oz.	250
Boiled egg 1.75 oz.	100
Ham, canned 1.25 oz.	94
Yeast extract 1/2 tsp.	86
Whole-grain bread 1 slice	72
Chapati, without fat 2 oz.	66
Potatoes, boiled 6 oz.	54
Oranges 1 medium	34
White bread 1 slice	33
Cabbage, boiled 3.3 oz.	25

MAGNESIUM

Magnesium is vital for calcium absorption. More than 50 percent of the body's magnesium is found in the bones. Although calcium is the central mineral in bone mineralization or calcification, the quality of the bone formed is dependent on magnesium. In the presence of magnesium, bones undergo a stronger calcification. This is because magnesium promotes and regulates parathyroid hormone, which stimulates bone cells to reabsorb calcified bone. Magnesium also helps regulate muscle cell function by aiding transmission of electrical impulses. When magnesium levels are low, muscles stiffen up or contract, causing cramps.

During stress, epinephrine prepares the body for "fight or flight." One of the effects of epinephrine is to take magnesium out of muscle cells and replace it with calcium. This gives muscles their rigidity for action. Because this magesium does not necessarily reenter the muscles once the stress is over, it is essential to maintain adequate magnesium levels.

Good sources of magnesium include nuts, seafood, green vegetables, brown rice, lentils, and split peas. It is important to avoid both alcohol and sugar; they increase magnesium excretion.

Zinc, copper, and manganese

Zinc, copper and manganese are important for bone health because they help form many of the constituents of new bone. Particularly good sources include shellfish, fish, nuts, and seeds.

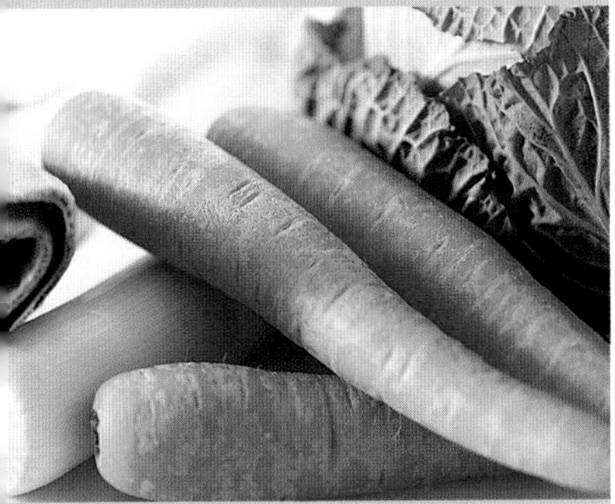

Protein

Adequate protein in the diet is essential. When protein is digested, it is broken down into amino acids, and these are used to build muscles, bones, and cartilage. Poultry and meat are good sources.

Recipes for healthy bones and muscles

Calcium, zinc, and vitamin D are the major nutrients needed for healthy bones and are available from a variety of food sources. Try these delicious and easy to prepare bone-friendly recipes to boost your intake.

STRAWBERRY AND BANANA SMOOTHIE

small container reduced-fat or soy strawberry yogurt
½ cup calcium-fortified orange juice
6 strawberries
medium banana

Blend all the ingredients together. Chill before serving.
Serves 1

CREAM OF ALMOND SOUP

1¼ cup low-fat milk
¾ cup reduced-fat fresh cream
grated zest of a lime
3.5 tbsp. reduced-fat spread (suitable for cooking)
3.5 tbsp. all-purpose flour
1 quart vegetable stock
onion salt
freshly ground pepper
⅓ cup ground almonds
pinch nutmeg

Combine the milk, cream, and lime zest in a pan; heat to just before boiling point, then remove from the heat. Melt the lowfat spread in a pan, stir in the flour, and add the stock gradually, stirring all the time, until the mixture thickens and boils. Season to taste with onion salt and ground pepper. Remove from heat.

Gradually stir the milk mixture into stock mixture until it is smooth and creamy, then add the ground almonds and nutmeg. Warm over low heat and cook for about 10 minutes.
Serves 4

Strawberry and banana smoothie
Per portion: 224 mg. calcium, 0.5 mg. zinc

Cream of almond soup
Per portion: 289 mg. calcium, 2.1 mg. zinc

CHILI KALE

I tbsp. olive oil
I clove garlic, crushed
I large white onion, chopped
2 lbs. kale, chopped and with stems removed
2 tsp. lime juice
I red chili, seeded and chopped
celery salt and black pepper to taste

Heat the oil in a pan, then add the garlic and onion. Sauté for about 10 minutes or until the onion is translucent. Add the kale to the pan and stir-fry for about 5 minutes. Stir in the lime juice and chili, season to taste, and serve immediately.
Serves 4

LENTIL AND STILTON LASAGNA

½ lb. red lentils
2 tbsp. olive oil
I clove garlic, crushed
I large onion, chopped
1.5 lbs. canned tomatoes, chopped
I tbsp. fresh rosemary
freshly ground black pepper
celery salt
8 ready-to-use lasagna noodles
I tbsp. reduced-fat spread (for cooking)
2 tbsp. all-purpose flour
I ¼ cups low-fat milk or soy milk
I cup blue Stilton, crumbled
I tbsp. fresh parsley

Cook the lentils in boiling water for 15 to 20 minutes, until tender, then drain. Heat the oil in a pan, add the garlic and onion and cook on low heat until translucent. Add the tomatoes, rosemary, and lentils; stir and bring to the boil. Season to taste with pepper and celery salt, reduce the heat, cover and simmer for about 5 minutes. Spread a small amount of the lentil mix in the bottom of a lasagna dish, then alternate layers of lasagna and lentil mix, ending with a lasagne noodle.

Melt the reduced-fat spread in a small pan, stir in the flour, and cook for a few seconds. Add the milk, a little at a time, and stir continuously to make a smooth white sauce. Remove from the heat and stir in the crumbled Stilton. Pour the sauce over the lasagna. Bake in a preheated oven (350°F) for about 40 minutes or until golden brown. Sprinkle with fresh parsley just before serving.
Serves 4

Chili kale
Per portion: 341 mg. calcium, 1.1 mg. zinc

Lentil and stilton lasagna
Per portion: 349 mg. calcium, 3.8 mg. zinc, 0.1 μg. vitamin D

Spinach and pepper salad with wild rice and feta
Per portion: 305 mg. calcium, 1.3 mg. zinc, 0.1 μg. vitamin D

Red Thai curry with tofu and mixed vegetables
Per portion: 584 mg. calcium, 1.8 mg. zinc

SPINACH AND PEPPER SALAD WITH WILD RICE AND FETA

1 cup wild rice
1 red pepper, sliced thinly
2 tbsp. feta cheese
2 scallions, chopped
1 tbsp. fresh basil, chopped
1 tbsp. fresh chervil, chopped
2 tbsp. toasted almonds
young leaf spinach
white pita bread, toasted

For the dressing

1 clove garlic, crushed
½ white onion, finely diced
3 tbsp. lime juice
1 tbsp. extra virgin olive oil
½ tsp. Dijon mustard

Cook the wild rice according to the instructions on the package, drain, and set aside to cool. In a serving bowl, mix together the cooked wild rice, red pepper, feta cheese, scallions, and herbs. Whisk the dressing ingredients together in a separate bowl. Pour the dressing over the rice mixture and serve on a bed of washed young leaf spinach, with the pita bread.
Serves 4

RED THAI CURRY WITH TOFU AND MIXED VEGETABLES

1 tbsp. olive oil
2 tbsp. red curry paste
1 or 2 fresh green chilies
¾ cup coconut milk
1 cup vegetable stock
1 large eggplant, diced
12 baby corn
½ cup snow peas
½ cup carrots, sliced
½ cup shiitake mushrooms, halved
1 green pepper, sliced
1 small can bamboo shoots, drained
1 lb. silken firm tofu, cut into 2-inch cubes and roasted as for Roasted tofu with tomato salsa, p. 67
1 tbsp. fish sauce
1 tbsp. honey
4 kaffir lime leaves
1 large handful of Thai basil leaves
handful of brazil nuts, toasted

For the red curry paste

6 dried red chilies
2 tbsp. chopped lemon grass
1 tbsp. chopped root ginger
1 tbsp. chopped spring onion
1 tbsp. chopped garlic
2 tsp. coriander seeds
1 tsp. cumin seeds
6 black peppercorns
1 tsp. salt
1 tsp. shrimp paste

Put the red curry paste ingredients in a food processor and blend until

Croustades with mixed toppings
Per portion: 433 mg. calcium, 1.9 mg. zinc, 2.1 μg. vitamin D

Sesame broccoli
Per portion: 91 mg. calcium, 0.9 mg. zinc

smooth. (This mixture can be stored for up to three weeks in a jar in a refrigerator.)

Heat the oil in a pan, fry the red curry paste and fresh chilies for one minute, then stir in 2 tbsp coconut milk (from the thickened part, which is at the top of the can) and cook, stirring constantly, for two minutes. Add the vegetable stock and bring to a boil. Toss in the eggplant, bring back to a boil and simmer for about five minutes. Add the remaining vegetables and cook for 5 to 10 minutes. Add the roasted tofu cubes and mix well. Stir in the fish sauce, honey, remaining coconut milk, and lime leaves, and simmer for another five minutes, stirring occasionally. Top with torn Thai basil leaves and toasted nuts. Serve with boiled rice.
Serves 4

CROUSTADES WITH MIXED TOPPINGS

1 white baguette, sliced and toasted
small can of sardines in tomato sauce
1 cucumber, cut into 1-cm cubes
low-fat spread
small tub reduced-fat soft cheese
handful dried apricots, chopped into small cubes
handful dried figs, chopped into small cubes

Sardine and cucumber topping
Mix the sardines with the cucumber cubes and divide between the toasted baguette slices.

Soft cheese and dried fruit topping
Spread the baguette slices with a small amount of low-fat spread. Top with soft cheese and dried fruit mix.
Serves 4

SESAME BROCCOLI

1 lb. broccoli florets
1 tsp. sesame oil
1 tbsp. dark soy sauce
1 clove garlic, crushed
1 tsp. toasted sesame seeds

Blanch the broccoli florets in boiling water for a couple of minutes; drain and place in a serving dish.

Make a dressing using the sesame oil, dark soy sauce, and crushed garlic, and pour over the broccoli.

Just before serving, sprinkle the sesame seeds over the top.
Serves 4

Eating for a healthy body weight

Maintaining a constant, healthy weight is beneficial for general well-being. By achieving and maintaining a good average weight, you should avoid the bone and muscle problems associated with being either over- or underweight.

COUNTERACTING WEIGHT GAIN

Weight levels are generally rising in the Western world: In 1999, 61 percent of American adults were classified as overweight or obese according to the BMI (body mass index), and in 2000, 38.8 million Americans were classified as obese. It is now known that being overweight puts you at a significantly higher risk for osteoarthritis and lower back pain, and places huge stress on the bones and muscles.

It can be easy to gain weight. You might be eating a balanced diet, without too many treats or high-calorie foods, and yet still be gaining weight. This is because unless you are burning as many calories in energy as you are eating as food, you will be storing the excess as fat. With today's typical lifestyle of driving to work, sitting at a desk for most of the day, and watching television most evenings, it is easy to see how this might happen.

It is not difficult to lose this excess weight, however. You simply need to redress the balance and burn more calories than you eat. However, rather than doing this just by eating less—a diet may help you lose weight but it may also deprive you of essential nutrients for bone and muscle health—it is better to combine a healthy eating style (such as cutting out foods rich in saturated fats) with an exercise program.

COUNTERACTING WEIGHT LOSS

At the other end of the scale, our perception of the ideal figure is driven by the media, with women in particular struggling to achieve lower weights and slimmer figures. Often this means turning to a crash diet or excluding important food groups from daily meals.

If taken too far, excess weight loss—including that associated with eating disorders—can be seriously detrimental to skeletal development (see page 69), and effort should be made to avoid this situation.

10 Ways to help maintain an ideal weight

Dietary and lifestyle factors play a part in maintaining body weight. If you are overweight, these measures may help you to lose a few pounds.

1. Switch to low-fat milk and low-fat spread instead of butter: Many low-fat dairy products are richer in calcium than their full-fat equivalents.
2. Choose leaner protein foods, such as chicken and fish, instead of red meats; venison is a low-fat red meat.
3. Snack on fresh or dried fruits and nuts rather than sweets and chips, both of which are high in saturated fat.
4. Grill, bake, or steam foods instead of frying or roasting. Trim off all visible fat before cooking.
5. Cut out sugar in tea and coffee, and cut out carbonated drinks. Sugar-free colas do not pile on weight, but they also do nothing to moderate a "sweet tooth."
6. Join in a ball game with the children and take every opportunity for family-friendly activities: swimming, skating, bowling, and so on.
7. Do the housework with a bit more vigor. Put on a CD to get you moving while you are ironing or dusting.
8. Walk to work if you can, or partway at least. Walk the children to school rather than taking the car—it's good for all of you.
9. Take a brisk walk during your lunch break. You will probably find at least one colleague who would be pleased to join you.
10. Walk up stairs instead of using the elevator; walk up escalators. Walk up and down the stairs at home frequently.

Body mass index (BMI)

The body mass index is a more reliable indicator of being over- or underweight than weight alone. Your BMI is your weight in kilograms divided by your height in meters squared.

- Below 20: underweight
- 20–25: acceptable weight range
- 25–30: clinically overweight
- 30–40: clinically obese
- 40+: morbidly obese

This chart represents the BMI scale visually so you don't have to do the arithmetic. Simply follow the line across from your weight until it meets your height to see which range you are in.

If you are overweight, devise a plan to lose weight slowly but steadily, and discuss it with your physician. By reducing your calorie intake to about 600 calories less than the average—an intake of about 1400 to 1500 calories for women and 1850 to 1950 for men—and taking up some form of exercise you should lose between 1 and 2 pounds a week.

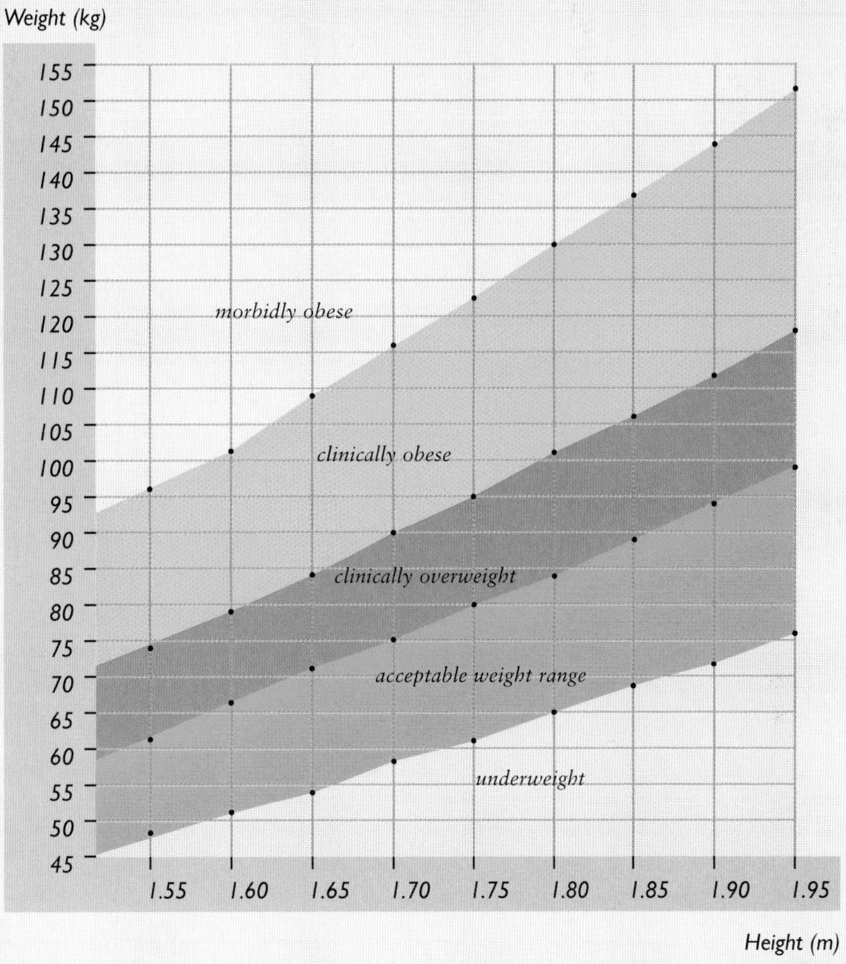

Weight (kg)
155
150
145
140
135
130
125
120
115
110
105
100
95
90
85
80
75
70
65
60
55
50
45
morbidly obese
clinically obese
clinically overweight
acceptable weight range
underweight
1.55
1.60
1.65
1.70
1.75
1.80
1.85
1.90
1.95
Height (m)

The best way to gain weight is to devise a diet of high-protein foods such as fish, lean meat, and dairy produce, combined with fiber-rich carbohydrates, such as bread and pasta, and foods rich in unsaturated fats, such as nuts and oily fish.

BLOOD SUGAR LEVELS

One of the most important ways of maintaining a stable body weight is to keep your blood sugar levels balanced. When your blood sugar level falls too fast or too low, you can become hungry, weak, and irritable. Often the first thing you reach for is refined sugar, in the form of a chocolate bar or pastry. We know that these processed foods give us a quick energy boost. The reality, however, is that our body is being set up for a roller-coaster ride of high blood sugar followed by a rapid decline. This leads to overeating and the storage of fat.

Glycemic index (GI)

The glycemic index is a measure of how quickly carbohydrates break down and release glucose into the bloodstream. Foods high on the index release glucose into the bloodstream quickly, causing blood sugar levels to soar. Foods with a low glycemic index provide a more steady supply of energy that lasts several hours. Thus, the GI can be a helpful tool for controlling sugar cravings and your weight. Foods that are high on the index include white rice, pasta, bread, cakes, and sweets, and peanuts. Foods that are low on the index include wholewheat or brown bread, pasta, and rice; almost all fruits and vegetables; and lentils and other legumes.

The best balance for stable blood sugar is 30 percent lean protein, 30 percent essential fat, and 40 percent low GI carbohydrates. Try to maintain this ratio at each meal and snack throughout the day.

Eat regularly

The most important factor in keeping blood sugar levels stable is to eat at regular intervals, ideally every 3 to 5 hours. Skipping meals or reducing your calorie intake does not help control weight in the long term.

Menopause and bone loss

Around menopause the rate of bone loss accelerates because of the fall in estrogen production. This is why women are more vulnerable to osteoporosis than men and need to find forms of estrogen to add to their diet.

NATURAL ESTROGENS

Isoflavones are a type of plant estrogen (phytestrogen) found in soy and are believed to mimic the effects of the natural female hormone estrogen, including the preservation of bone mass. Some studies have suggested that soy-rich diets and soy supplements can help alleviate some symptoms of menopause, including osteoporosis, hot flashes, and depression.

A study published in the journal *Obstetrics and Gynaecology* in 2001 looked at the dietary intake of isoflavones in more than 470 postmenopausal Japanese women. The findings suggest that a diet rich in soy may help women retain stronger bones after menopause, thus reducing their risk of osteoporosis. In this study, women in early and late stages of menopause who consumed the highest amount of isoflavones in foods such as tofu, boiled soy beans, and soy milk had significantly denser bones than those who consumed the lowest levels of isoflavones. Isoflavones did not, however, have an effect on menopausal symptoms in the late stages of menopause. This research suggests that high intakes of soy products are associated with an increase in bone density in postmenopausal women.

What about soy supplements?

According to results from a small study published in the *Journal of the American College of Nutrition* in 2002, isoflavone-rich soy supplements do not appear to boost bone mass in young women. This suggests that the benefits in bone health associated with isoflavones in some studies of older women do not extend to young women. The researchers concluded that the high levels of natural estrogen produced by young menstruating women overshadow any effect isoflavones from soy might have on bone.

Calcium supplements

There is some debate as to whether taking calcium supplements during menopause helps prevent the associated bone loss. Studies that have looked at supplementing the diets of perimenopausal women—those in the five years before menopause—with calcium have shown little effect on slowing down the rate of bone loss.
It is generally accepted by health professionals

SAMPLE MENU *DAILY DIET PLAN TO PREVENT OSTEOPOROSIS*

breakfast

- *1 bowl muesli topped with a handful of dried figs or dried apricots, served with low-fat milk or calcium-fortified soy milk*
- *1 glass calcium-fortified orange juice*
- *1 piece of toasted white bread topped with reduced-fat spread and fruit preserve*

lunch

- *Toasted pita bread stuffed with mozzarella cheese, basil, and tomato*
- *1 container low-fat yogurt*

dinner

- *1 serving of red Thai curry with tofu and mixed vegetables (see page 62)*
- *1 bowl of boiled rice topped with strips of toasted seaweed*
- *1 serving of mixed fruit salad with low-fat fresh cream*

snacks and drinks

- *1 orange*
- *Water (either tap water in a hard-water area or calcium-fortified bottled water)*
- *1 scone with reduced-fat spread*

that hormone replacement therapy is the most effective way of preventing the loss of bone during this period.

However, there is strong evidence from at least 20 scientific studies that calcium supplementation does prevent bone loss in postmenopausal women. These studies have found that women who did not take extra calcium lost around 1 percent of bone mass per year, compared with those taking calcium, who did not suffer any bone loss. There is also good evidence to suggest that supplementing with calcium may prevent fractures.

ROASTED TOFU WITH SALSA SAUCE

1½ tbsp. olive oil
14 oz. silken firm tofu, cut into 1-inch cubes

For the salsa

8.5 oz. fresh tomatoes, skinned and diced
3.5 oz. cucumber, diced
2 chili peppers, sliced
2 tbsp. scallions, thinly sliced
½ clove garlic, crushed
½ tsp. sugar
½ tsp. tarragon
1 tbsp. lime juice
2 tbsp. fresh coriander, chopped
freshly ground black pepper

Preheat the oven to its highest setting. Pour the olive oil on a large foil-lined roasting pan, spreading it evenly. Space the tofu cubes on the oiled foil, and place the roasting pan on the top shelf of the oven and roast for about 15 to 20 minutes or until the cubes are crisp and golden brown.

Prepare the tomato salsa by combining all the ingredients together in a bowl. Mix well and set aside. Take the roasted tofu cubes out of the roasting tin and place on to a serving dish. Spoon over the prepared salsa.

Serves 4

Versatile tofu

In this colorful variation on the recipe above, steamed baby corn, mixed peppers, and snow peas have been added to the roasted tofu, and the dish is served with rice noodles.

8 Great ways to include more soy in your diet

For menopausal women who do not want to begin hormone replacement therapy (HRT) or for those for whom it is unsuitable, boosting soy intake may be beneficial.

1. Drink a glass of sweetened soy milk instead of your usual coffee or tea at breakfast.
2. Top your bowl of muesli with a container of soy yogurt instead of milk. Soy yogurt is available in many fruit flavors.
3. For a tofu sandwich spread, mix 9 oz. tofu, 2.5 oz. red onion, 0.75 oz. soy sauce, 2 oz. carrots, 1 tsp. paprika, 2 oz. coriander, and 4.5 oz low-fat cream in a blender.
4. Eat burgen bread (containing soy and linseeds) instead of your normal whole-grain or white bread.
5. Stir-fry tofu, vegetables, and noodles for a quick lunch or supper dish, or roast and serve with a colorful salsa or salad.
6. Add sprouted soy beans to salads and garnishes for extra crunch.
7. Add canned soy beans to stews, casseroles, and soups.
8. Try this tofu dressing instead of salad dressing: Mash one package soft tofu with 1 tbsp. sugar, 1 tsp. salt, 1 tsp. soy sauce, 1 tbsp. rice wine vinegar or cider vinegar.

Although soy foods are the most popular source of phytestrogens, many other foods are rich sources of these compounds, including:

- alfalfa sprouts;
- legumes;
- wheat;
- licorice;
- fennel;
- seed oils, particularly flaxseed;
- celery;
- green beans; and
- pomegranate seeds.

Plants contain many types of phytestrogens, as well as minerals and other constituents that help womens' bodies to modify phytestrogens. Red clover, for example, contains all four major types of phytestrogen: lignans, coumestans, isoflavones, and resorcylic acid lactones.

Bioflavonoids

It has been suggested that foods containing bioflavonoids can also benefit menopausal women, because bioflavonoids bear a structural resemblance to a form of estrogen. In addition, bioflavonoids have been shown in numerous studies to reduce hot flashes. Foods that naturally contain bioflavonoids include oranges, grapefruit, tangerines, peppers, linseed, and garlic.

THE NEED FOR HEALTHY, ACTIVE GUT FLORA

Plant hormones, including most phytestrogens, cannot be directly absorbed by humans. However, we can convert them into a form we can use with the help of our digestive bacteria. When we take antibiotics, the absorption of phytestrogens plummets because the good bacteria are killed off along with the bad. It is therefore important to eat more bio-yogurts or probiotic dietary supplements, which contain good bacteria such as Acidophilus, Bifidus, and Bugaricus, when taking these drugs.

Digestion has particular relevance for women during menopause, because both digestion and subsequent absorption are vital for the prevention of osteoporosis, and indeed, many of the major degenerative disorders.

The uptake of nutrients, particularly calcium and magnesium, tends to decline with age, because of the general state of the digestive system, which tends to become sluggish. Bitter foods are helpful; they can enhance the function of the upper digestive tract and improve the assimilation of nutrients. Foods generally classed as "bitter" include endive, chicory, mustard greens, grapefruit, and the outer leaves of artichokes.

Osteoporatic and associated fractures in hospitals and nursing homes cost Americans $17 billion in 2001—$47 million per day.

Eating disorders and bone loss

By altering estrogen production, eating disorders—typically anorexia nervosa and bulimia nervosa—can lead to serious nutritional deficiencies and long-term health problems, including osteoporosis.

Anorexia nervosa and bulimia nervosa are the most prevalent eating disorders in society today. Often linked to intense anxiety over appearance and body weight, they are characterized primarily by an intentional forced starvation in the case of anorexia and episodes of purging in the case of bulimia. Both disorders can have serious implications for bone health and affect the normal functioning of all parts of the body.

Eating disorders can cause bone loss within six months. Of the eating disorders, anorexia has the most profound effect on bone health. A study published in the *Journal of the American Medical Association* in 1996 found that 50 percent of young patients with anorexia were also diagnosed as suffering from premature osteoporosis.

EFFECTS OF EATING DISORDERS ON BONES

- Poor nutritional status The use of laxatives and diuretics associated with bulimia nervosa causes alternating bouts of constipation and diarrhea—a process that has a detrimental effect on the absorption of nutrients, such as calcium, from the diet. This can lead to a significantly higher risk of developing osteoporosis.
- Low weight People suffering from anorexia will be as much as 20 percent below their normal body weight. Being underweight is a risk factor for osteoporosis in men as well as women—bone density decreases as body weight falls. The more severe the weight loss, the more bone is lost.
- Menstrual dysfunction If a woman's body reaches an extremely low weight, it is likely that menstruation will become irregular and may even cease altogether (amenorrhea). The sex hormones, estrogen in particular, are important in the formation and maintenance of bone mineral. Menstrual problems result in estrogen deficiency, so this will have a detrimental effect on the bones and increase the risk of osteoporosis later.

EARLY ACTION

If left untreated or undetected, eating disorders and amenorrhea can lead to compromised bone health for life. Prompt diagnosis can lead to early treatment and a better prognosis. Warning signs of an eating disorder include

- precoccupation with food;
- eating small portions;
- increasing criticism of body weight;
- over-exercising;
- cutting out certain foods;
- skipping meals and fasting;
- abusing diet pills, laxatives, and diuretics;
- self-induced vomiting; and
- wearing bulky, oversized clothing.

Women, most commonly 15 to 25 year olds, are more likely than men to be affected by an eating disorder.

FOR TEENAGERS

Bone loss early in life

Teenage eating disorders are particularly serious because they coincide with a major period of bone formation. About 40 percent of skeletal calcium should be established during the teenage years. However, because of the severe reduction of food intake, the body becomes deprived of the calcium it needs to build proper bone mass. Teenagers with eating disorders are losing bone mineral when they should be gaining it, so it is unlikely they will achieve maximum bone density. This can lead to serious bone problems later in life.

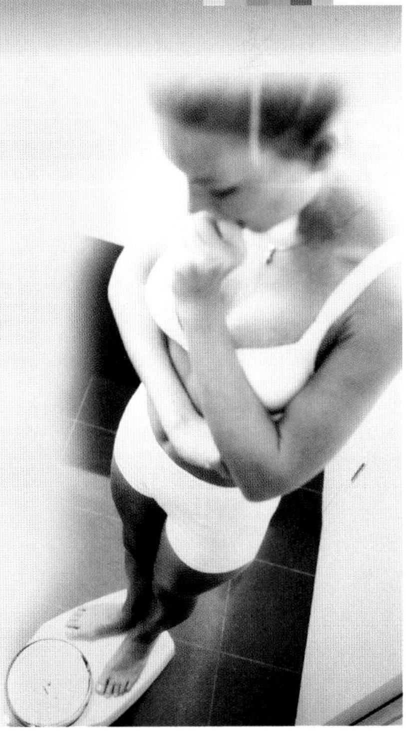

Cutting down on toxins

Smoking and drinking more alcohol than the recommended limit can both have serious effects on bone density. These effects appear to be dose-related: The more an individual smokes or drinks, the worse it is for the bones.

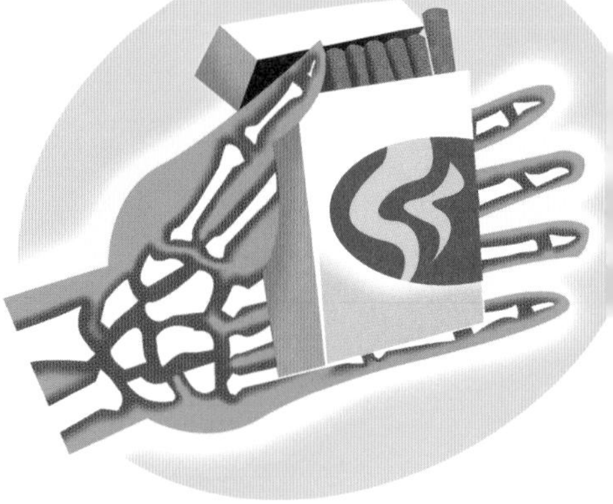

Eating a healthy, balanced diet and getting regular weight-bearing exercise are two vital factors in building bone density in the middle years and preserving it through menopause and beyond. A third is cutting down on toxins: Excess tobacco and alcohol are both harmful for the bones.

ALCOHOL AND BONES

Moderate drinking is not detrimental to bone health; indeed a study conducted in 1999 demonstrated that women who consumed 2.5 oz. of alcohol a week (the equivalent of a glass of wine a day) had higher bone density of the lumbar spine than nondrinkers. High alcohol intake, however, increases the rate at which bone is lost and replaced. For general health as well as bone health, it is important to limit yourself to no more than three to four units a day if you're a man and two to three units a day if you're a woman and to have one or two alcohol-free days a week. One unit of alcohol is generally defined as

- a glass of wine,
- a standard measure of liquor,
- 8 ounces of ordinary (normal strength) beer, lager, or hard cider.

Excess alcohol may also lead to unsteadiness, which contributes to falls and fractures and is particularly dangerous for those at risk of osteoporosis.

Alcohol and the joints

Alcohol also contributes to gout, a painful condition that occurs when the body either produces excess uric acid or is unable to process it effectively. Crystals of uric acid are deposited in the joints, usually the joint of the big toe, causing redness, inflammation, and pain.

Beer is thought to be particularly harmful because it contains purines. Purines are compounds that are also found in organ meats such as liver, dried peas, and oily fish. Anyone who has suffered from gout is advised not to drink alcohol regularly. Moderate infrequent drinking is believed to be harmless.

Moderate consumption of alcohol may increase bone density by speeding the conversion of testosterone into estradiol, a hormone that prevents bone loss.

SMOKING AND BONE LOSS

Smoking depresses estrogen production, robbing the bones of the hormone's protective effects in premenopausal women. Smoking 20 cigarettes a day leads to a reduction of between 5 and 10 percent in the amount of bone, and nearly doubles the risk of fracture in later life.

Menopause may occur up to two years earlier in female smokers than in women who do not smoke. This exposes them to the risks of bone fragility earlier.

THE DUAL THREATS

Both smoking and excess alcohol intake may suppress the appetite, with the result that the nutritional status of smokers and drinkers may be compromised. Alcohol disrupts the digestive process. A lack of dietary calcium, or its malabsorption, will contribute to bone loss.

Alcohol, smoking, and street drugs may all interact with prescribed medication, including HRT for those taking it to preserve bone density.

For the sake of your health and quality of life in the middle and later years, it is important to use every available resource to help you to quit smoking and bring alcohol drinking within advised limits.

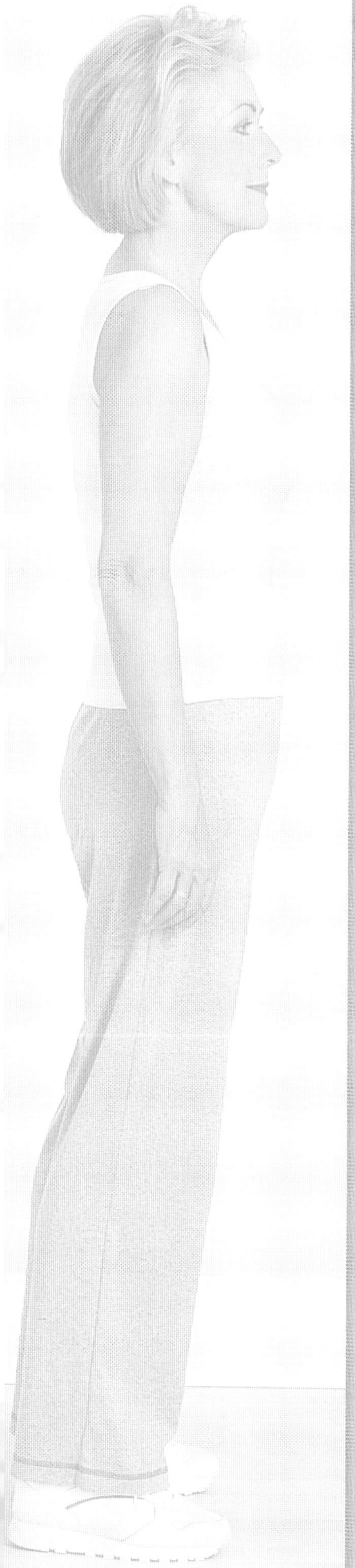

A BONE- AND MUSCLE-HEALTHY LIFESTYLE

Although generally robust, the musculoskeletal system is easy to strain as you go about your daily routine. However, keeping it healthy is not difficult as long as you are aware of potential risk factors and can act appropriately to avoid the worst of the pitfalls.

72

How you hold your body in your daily life can have an enormous impact on how your bones, muscles, and joints stand up to wear and tear.

76

Sports and play are good for adults and children, but injuries to the muscles and bones are common. Taking simple precautions should avoid problems.

79

Some workplaces are more hazardous for the bones and muscles than others, but there is a lot we can all do to prevent accidents.

82

A good night's sleep relaxes tired bones and muscles, so they are fit for the day ahead. Your choice of bed has a major impact on how well you sleep.

83

The fit of your shoes has implications not just for your feet, but for your legs and spine, too. Well-fitting shoes during childhood will bring rewards.

The importance of posture

As we get older and spend more time sitting, we tend to let our posture go. At the very least, this can restrict blood circulation and result in minor injuries, but it also contributes to back pain, falls, and fractures.

Having good posture means holding your body in its correct alignment, whether you are standing or sitting, and this is primarily achieved by maintaining the optimum function of your muscles. The major muscles of the body are the prime movers of your bones, but the lesser muscles work to keep your body in its correct alignment—with one section stacked above another—and maintain the body's natural curves.

Poor standing posture is characterized by rounded shoulders, a head tilted forward with the chin down or jutted out, a rounded upper back and arched lower back, protruding buttocks or tummy, and a flattened chest.

STANDING POSTURE CHECK

- Stand with your back against a wall with equal weight on both feet and your feet hip-width apart.
- Breathe and relax into this position for a few moments.
- Press the back of your neck toward the wall so that your chin drops and the back of your neck stretches.
- Relax the lower back and feel your shoulders drop behind you.
- Lift your ribcage to lengthen your torso.
- Relax your neck and feel a lift from the top of your head to the ceiling.
- Take a step away from the wall.
- Press the small of your back in toward the wall so that the coccyx (tail bone) drops toward the floor.
- Practice this regularly until perfect posture becomes instinctive.

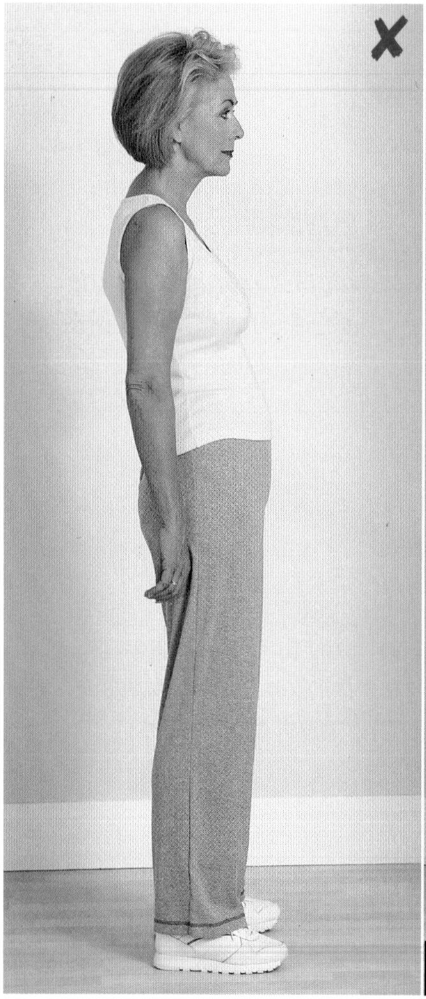

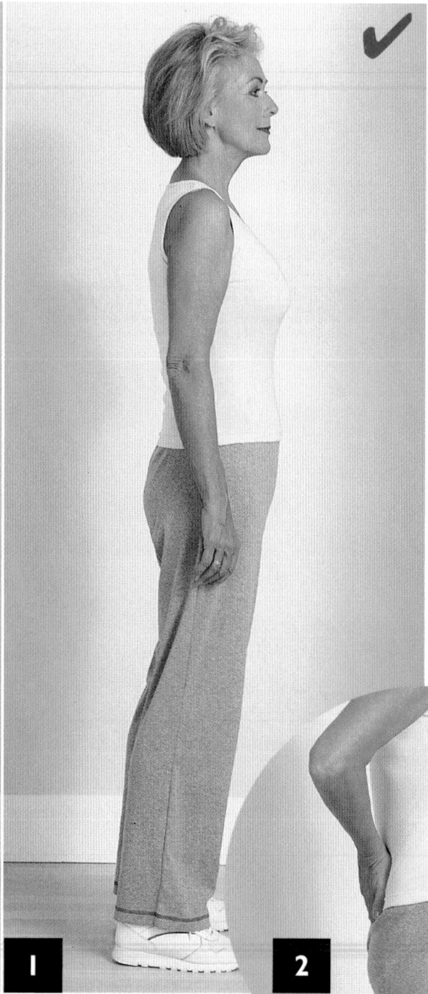
1

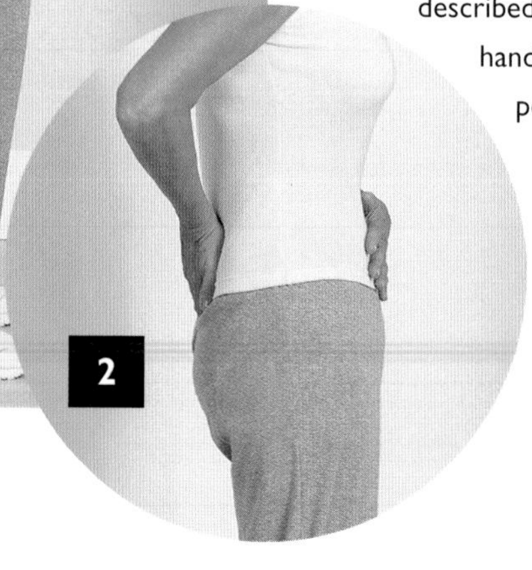
2

How to stand

Poor posture (far left) compromises the health of the spine and contributes to back pain. Classic poor posture is to stand with sagging abdominals and rounded shoulders, causing the neck and head to sag, and making you look shorter.

1 Good posture begins by standing with your feet hip-width apart and toes turned out slightly. Avoid "locking" your knees. Keep your hips square to the front and tilt your pelvis (below). Keep your abdominal muscles pulled in and lift your ribcage. Keep your back straight and shoulders back, with your head straight and chin parallel to the floor.

2 Tilting your pelvis ensures that your spine is correctly aligned. Stand as described above and place your hands as shown. Tilt your pelvis so that your hips move forward and up. Tighten your abdominal muscles to keep the pelvis in this correct position.

WALKING AND JOGGING

Everyone has a different style when walking or running. Some of us take short strides, others long. Some have a high knee lift, but for others it is low. Some of us land on our heels and others our toes.

When it comes to posture, however, there are a few key points to remember whether you are out for a leisurely walk or a jog: The most important is that proper alignment offers maximum power with minimum effort. By keeping your upper body upright and relaxed, you avoid wasting muscle energy in trying to keep your balance in check. By leaning either excessively forward or back while jogging, you put unnecessary stress on your hips, knees, and back, and this will lead to injury of one kind or another.

The benefits of being relaxed

Focus your muscle power on moving forward: Do not waste energy in clenching face muscles or fists. Your arms should be relaxed. Use them for balance and coordination, and do not raise them higher than chest level or in front of your body as you run—keep them at your sides.

The benefits of an upright posture

Maintaining an upright upper body when running or walking brings immediate benefits:

- It will be easier to increase your speed, as your legs are in the correct position to extend fully, with little extra effort.
- You will sustain fewer injuries.
- You will breathe easier: Correct posture allows all your organs to sit in the space intended for them.
- There is less likelihood of developing a stitch.

RUNNING IN STYLE
Head and shoulders Keep your head high, maintain a relaxed expression, and relax your shoulders.
Keep upright Lift your ribcage and hold your abdominals in to maintain an erect posture.
Arms and hands Swing your arms easily; relax your hands and gently curl your fingers.
Hips and pelvis Tilt your pelvis forward and keep your hips to the front.
Legs and feet Push off from the balls of your feet, bending through your knees.

HOW TO SIT

Poor seated posture can cause physical deformity and pain. When you sit with poor posture,

- the lower back slumps and eventually starts to ache;
- the upper back curve becomes exaggerated;
- the neck is then compacted as the head pokes forward, leading to a rounded back and shoulders; and
- stomach muscles are lax.

Behind the wheel

Years of sitting badly when driving will take its toll, particularly as bones get weaker. Have the seat upright and keep your lower back pressed to the back of the seat to maintain an upright posture. Constantly pressing on the pedals transmits large forces up the legs to the lumbar spine.

Seated posture check

These pointers apply whether you are working at your desk, relaxing on a couch, or driving.

- Aim to keep the top of your head pressing toward the ceiling and your chin parallel to the floor.
- Press your shoulders toward the back of the chair.
- Lift your ribcage off the waist so that you sit tall.
- Try to sit with the base of your spine against the back of the chair.
- Pull the stomach muscles in to support your torso, keeping your back straight and not bowing out to the back.
- Avoid crossing your legs, which restricts circulation, but sit with your legs in a parallel position.

HOW TO LIFT

Lifting an object of any kind is a potential risk for all people but particularly for those with some degree of bone or muscle weakness. Lifting incorrectly can cause back and hip injuries, so any lifting work should be approached with caution.

Assessing the task

Always think carefully about what you are going to lift. Be sensible: Can you really lift it alone? Can you hold it close to your body? Can it be broken down into smaller loads? Could you use a trolley or cart to move the object? Is the object so large that it will restrict your vision? If it is, don't lift it. Only attempt to lift objects that you are certain you can manage on your own.

If you are lifting an object with someone else, be sure you both know when to make a move: An object that needs two or more people can harm a single person who tries to lift at the wrong time. Be sure you both know when to put a load down, as well.

Picking up the load

- Tuck your pelvis in and contract the abdominal muscles to support your torso.
- Squat, bending your knees and keeping your heels on the floor. Your back should be straight, extending forward from the hips.
- Hug the load: Try to bring the object you are lifting as close to your body as possible.
- Check that your abdominal muscles and the pull on your pelvic floor are still contracted.
- Press down through your legs and buttocks to bring yourself to standing. Try to keep the movement smooth.

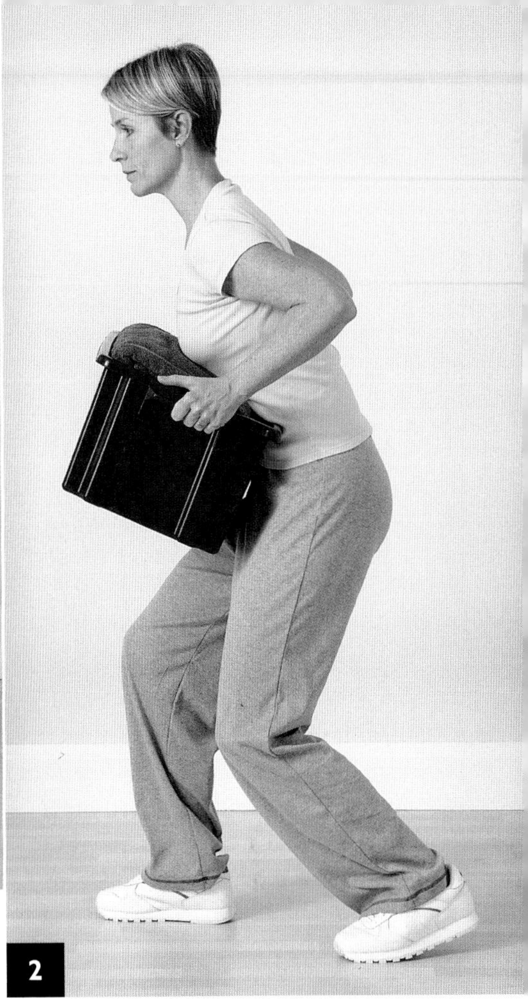

Picking up a load

1 Stand with your feet hip-width apart and one leg slightly in front of the other as close to the load as you can. Bend your knees and keep your back straight. Tilt your pelvis and pull your abdominals in.

2 Take hold of the weight and bring it in toward your body. Stand up slowly, keeping your back upright so that your legs are doing the work and you get maximum leverage with minimum strain.

You can lift almost twice the weight if it is at waist level than you can when it is at foot or head level.

Putting the load down

- Face the spot you have chosen and lower the load slowly.
- Bend your knees and let your legs, not your back, do the work.
- Keep your fingers away from the bottom of the load, and place it on the edge of the surface, then slide it into place.

SAFE LIFTING

When getting heavy loads in or out of the trunk of your car, stand with your feet shoulder-width apart, bend your knees and start to squat, bending at your hip joints rather than at your waist. Tighten your abdominal muscles as you lift or lower an object into the trunk.

If you have a shopping cart, always push rather than pull. You have twice as much power and less chance of injury.

When picking up children, drop down on one knee and pull the child in toward your chest. Then push back to a standing position with your legs. Baby carriers distribute a baby's weight and so should not place undue strain on your back, but be guided by your baby's weight.

If you have to lift an object that's above shoulder level, use a stool or stepladder to avoid over-reaching.

positive health tips

Avoid injury when lifting

- Don't twist at all while lifting or carrying heavy loads.
- Don't jerk or move suddenly while lifting.
- Don't lift above your head or over an obstruction.
- Don't lift objects that are slippery or wet. Make sure you have a good grip at all times.
- Don't lift objects that obscure your vision or footing.
- Don't bend at the waist. Keep your abdominal muscles contracted and bend from your hips and at your knees.

Sports and play safety

Sports promote physical fitness, boost self-esteem, and foster a feeling of well-being, and children learn about the world through play. However, neither activity is without risk to the bones and muscles of both children and young adults.

Accidents are the most frequent cause of death and permanent injury or disfigurement in young people. Although injuries in young people often heal well and faster than in older people, any disfigurement or incapacity will remain with a young person for a larger proportion of life. It may have a greater impact on lifestyle and is likely to result in more limitation than it would in someone older. Prevention is always the best option, and awareness of risk factors is a good starting point.

PLAYGROUND SAFETY

In 2001, doctors and hospitals treated more than 527,000 injuries caused by or related to playground equipment, at a cost of more than $11 billion. Keep the following factors in mind when assessing playground safety:

- The surface should be even and child-friendly—either rubberized slabs or bark chips; packed earth and concrete are too hard.
- The design of equipment such as swings and slides should be appropriate for a child's age and activity.
- There should be a specific "use" zone marked around the equipment: A small child cannot be expected to stop a swing, for example, if another child "strays" into an unmarked use zone. Safe distances should be clearly marked.
- The playground should be laid out so that activities in one area do not interfere with those in another.

The playground offers unmatched possibilities for the development of perception and physical skills, including motor control, in the forms of running, climbing, dodging, swinging, and throwing. Children of different ages and sizes should have separate areas, because equipment may need to be sized differently. Regular maintenance is vital: All equipment should be checked for rot, loose fittings, and rusting.

Risk assessment

Some things are relatively easy to assess. Children should be able to grasp handrails easily, steps and rungs should be evenly spaced, elevated surfaces should have adequate barriers, platforms should be safe at the top, swings should be attached and hang properly, and stairways should have handrails that are easy for small hands to grasp. Falls are the most common playground injury, which may be compounded by children hitting an unyielding piece of equipment during the fall.

Children should be dressed for comfort: They need elasticity and agility to deal with physical challenges. Avoid scarves and drawstrings: They may catch on a piece of equipment and endanger a child. Most of these issues can be quickly assessed and a child can be told to avoid a certain part of a playground if necessary.

SAFE SPORTS FOR CHILDREN

Sports activities for younger children and young athletes are more than just play. They aid in physical and psychological development and improve physical fitness, coordination, and self-discipline. They offer valuable opportunities to learn teamwork and to get acquainted with the intricacies of winning and losing. Whereas general safety precautions such as using equipment wisely, wearing appropriate clothing, and warming up apply, sports may need modifying to be suitable for children. If in doubt, ask a qualified coach.

Children should never be pressured to play sports or train: A child faced with unrealistic expectations may ignore the warning signs of overexertion and continue to train through pain, causing potentially lifelong problems. Encourage an atmosphere in which sports are fun.

Children's bones are softer and more elastic than adults', their ligaments are more malleable, and their muscles are capable of greater stretching, but their strength has not yet reached its maximum. Children still have growth plates, specialized structures at the ends of long bones responsible for longitudinal growth. Injury to these plates can jeopardize growth, leading to deformity later. Generally, the younger a child is injured, the greater the degree of

Climbing equipment is potentially the most dangerous playground equipment, causing almost 200,000 of the 527,000 playground injuries in people under 20 years old in 2001.

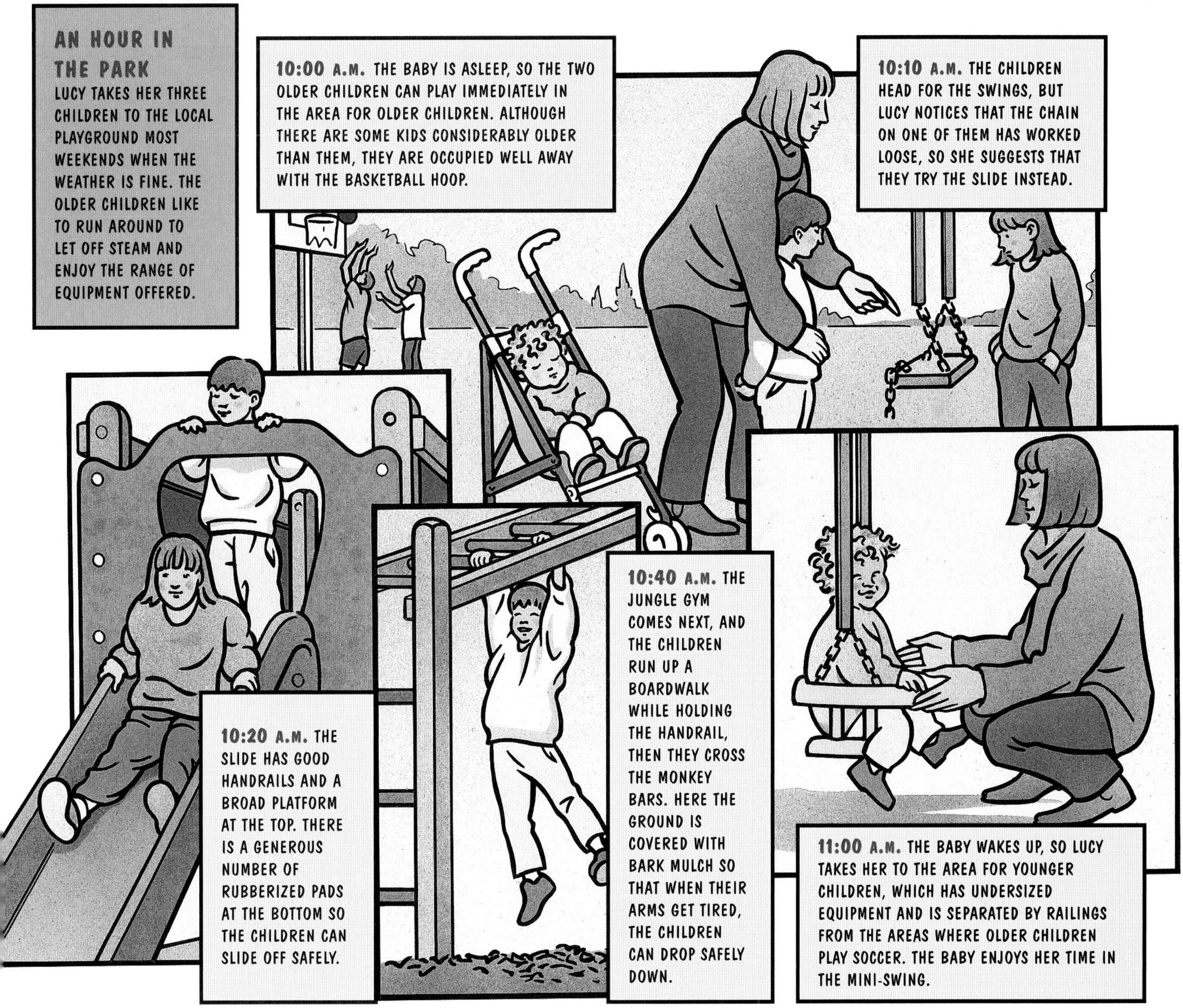
AN HOUR IN THE PARK
LUCY TAKES HER THREE CHILDREN TO THE LOCAL PLAYGROUND MOST WEEKENDS WHEN THE WEATHER IS FINE. THE OLDER CHILDREN LIKE TO RUN AROUND TO LET OFF STEAM AND ENJOY THE RANGE OF EQUIPMENT OFFERED.
10:00 A.M. THE BABY IS ASLEEP, SO THE TWO OLDER CHILDREN CAN PLAY IMMEDIATELY IN THE AREA FOR OLDER CHILDREN. ALTHOUGH THERE ARE SOME KIDS CONSIDERABLY OLDER THAN THEM, THEY ARE OCCUPIED WELL AWAY WITH THE BASKETBALL HOOP.
10:10 A.M. THE CHILDREN HEAD FOR THE SWINGS, BUT LUCY NOTICES THAT THE CHAIN ON ONE OF THEM HAS WORKED LOOSE, SO SHE SUGGESTS THAT THEY TRY THE SLIDE INSTEAD.
10:20 A.M. THE SLIDE HAS GOOD HANDRAILS AND A BROAD PLATFORM AT THE TOP. THERE IS A GENEROUS NUMBER OF RUBBERIZED PADS AT THE BOTTOM SO THE CHILDREN CAN SLIDE OFF SAFELY.
10:40 A.M. THE JUNGLE GYM COMES NEXT, AND THE CHILDREN RUN UP A BOARDWALK WHILE HOLDING THE HANDRAIL, THEN THEY CROSS THE MONKEY BARS. HERE THE GROUND IS COVERED WITH BARK MULCH SO THAT WHEN THEIR ARMS GET TIRED, THE CHILDREN CAN DROP SAFELY DOWN.
11:00 A.M. THE BABY WAKES UP, SO LUCY TAKES HER TO THE AREA FOR YOUNGER CHILDREN, WHICH HAS UNDERSIZED EQUIPMENT AND IS SEPARATED BY RAILINGS FROM THE AREAS WHERE OLDER CHILDREN PLAY SOCCER. THE BABY ENJOYS HER TIME IN THE MINI-SWING.

deformity. Most injuries, however, are minor and can be treated so that growth is not impaired.

Most sports-related injuries in children are caused by contact sports, but for all sports and games simple preventive measures can make the activity safer:

- Divide children into groups of equivalent size and ability: Grouping by age may not be sufficient, because children mature at very different rates.
- Avoid overly rigorous training, especially for small children.
- Make sure there is always well-maintained safety equipment.

Children who experience any of the following problems should be seen by a specialist:

- Inability to play following an acute or sudden injury.
- Decreased ability to play because of chronic or long-term complications following an injury.
- Visible deformity of the arms or legs or severe pain that prevents the use of an arm or leg.

Initial treatment for a simple injury is RICE (see page 95). Wash any small wound in water and antiseptic solution. If you suspect a sprain (if the child is in acute pain and has difficulty holding a wrist or arm or putting weight on a foot or leg), take the child to the nearest emergency room. A more serious injury, such as an open fracture, requires immediate and specialist transportion to a hospital. If you are supervising a group of children training or playing, it is useful to know the basics of first aid so that you can provide adequate help until the paramedics arrive.

Playing it safe

The benefits of exercise for everyone are undeniable, and introducing children early to its advantages pays dividends. Weight is becoming a health problem for American teenagers; 13% of children and adolescents are classified as overweight, with an increased risk of being overweight as adults and developing diabetes and cardiovascular problems. But children do need to take care when they exercise: Their bones and muscles are undeveloped compared with adults', and sprains and fractures are easy to acquire.

1 Shoe shop

The wrong shoes on the wrong surface can harm a child's feet, ankles, knees, and spine. A general-purpose shoe is fine for playing many different sports, but anyone seriously training and playing several times a week needs sports-specific shoes.

2 Get padded up

In-line skating is great fun, but falls are common, especially as a youngster is learning the technique. You can minimize the risk of broken bones by wearing knee and elbow pads, as well as a helmet. Helmets are also vital when bicycling.

3 Play and stretch

It is as important for children and teenagers to warm up and stretch as it is for adults. Young muscles are easily harmed, so any training session or match should include an adequate warm-up and cool-down.

4 Baby gym

From toddlerhood onward, children enjoy play gyms. Floors and equipment are usually very safe for children, but an awkward fall or twist can still lead to sprains and strains. Supervision is the key: Check the child-to-adult ratio. Child-to-adult ratio requirements for daycares vary by state.

5 Safe bouncing

It is fairly easy for an excited child to fall in a moon bounce and sprain a wrist. The likelihood increases if too many children of different sizes are allowed on at the same time: The older ones can push the little ones off. Similarly, check that trampolines are not overloaded when children play.

Health and safety in the workplace

How you sit and stand at work can mean the difference between a productive day and long-term pain or serious disability. Back pain in particular causes more lost working days than any other problem, but it can be avoided.

Work-related injuries cost Americans more than $120 billion per year.

Anyone whose work involves repetitive hand, finger, or arm movements; upper body twisting; reaching; hammering; pushing and pulling; or lifting is at risk of an upper limb disorder, or repetitive strain injury (RSI), characterized by swelling, aches and pain, and difficulty moving. RSI is commonly linked to office workers, but it can occur in anyone whose job involves repetition or force in an awkward posture (including working in a confined space). Assembly line workers, for example, are at risk, as are packers, food handlers, mechanics, and construction workers.

It is the responsibility of both employers and employees to monitor workplace safety in order to minimize the chance or impact of such conditions. This could involve, for example,

- redesigning work stations, perhaps moving components closer to workers to cut down on reaching;
- replacing hand tools with quality industrial tools;
- phasing out piecework so that employees are not stressed to achieve unreasonable targets;
- more flexible working patterns, with a greater number of shorter breaks;
- training in skills and posture for at-risk employees;
- improved heating, lighting, and ventilation; and
- adequate protective clothing.

BEATING BACK PAIN

One of the most important things you can do to prevent back pain is to keep yourself physically fit, strong, and flexible. Strong stomach and back muscles reduce the likelihood of back pain, protect against back injury, and speed healing after an injury or back surgery.

Bear in mind that the body can only tolerate sitting or standing in the same position for about 20 minutes: Holding the same position causes stress and discomfort. Move frequently: Stretch, stand, or walk if you are involved in a sedentary occupation, and walk or shift your weight if you have to stand for long periods. This will restore elasticity to the tissues protecting the joints. Keep in mind, too, that fatigue makes you move more awkwardly, which can also stress your back. If you are frequently tired at work, take steps to ensure that you get enough good, restful sleep.

If you do suffer from back pain, evidence now suggests that resting until the pain has gone can be harmful. It is better to continue normal activity, including going to work. Stopping all activity in the hopes of avoiding the pain is not a good idea. If you think work is causing the problem, ask if your tasks can be modified until your back pain has healed.

SLIPS AND FALLS

A fall or slip in the workplace can lead to pulled muscles or broken bones. In severe cases, bad falls can even result in amputation of a limb. According to the Bureau of Labor

Back to front
The key to a pain-free back is strong abdominal muscles that stabilize and support the torso. Exercise balls can help work the back and stomach muscles in tandem.

BE SAFE AT WORK

Workplace safety is the responsibility of everyone, and even high-risk environments can be safe if employers and employees follow all safety guidelines.

STAIR SENSE

Use handrails wisely, and avoid carrying heavy loads, especially if they obscure your view. Don't run up and down: The risk of missing your footing is too great.

LADDER WORK

Ladders should be the correct length for the task at hand and located on a stable surface. Don't extend your normal reach—always get down and move the ladder.

MOUSE WISE

Take frequent breaks from using a mouse and stretch your fingers. Remove your hand from the mouse when you are not using it rather than resting it on the mouse.

Statistics, cases of occupational illness and injury in the private sector have fallen steadily since 2001.

Falls are most common in jobs involving lifting, ladders, and scaffolding, and they usually occur when the work environment has changed in some way—a dry floor gets wet or oily or an obstruction is placed in an unaccustomed area, for example. The key is to be alert so you can spot potential hazards and deal with them before they cause a problem for you or anyone else.

UNEXPECTED HAZARDS

Walk only where you are supposed to rather than taking a shortcut through machinery or storage areas, and always watch where you are going—don't get distracted by talking to a colleague. On stairs, use any handrails and avoid carrying loads with both hands, especially if your vision is blocked. Make two trips rather than compromising the safety of your back or risking a fall by carrying one heavy load.

Work-related back injuries are the number one occupational hazard in the United States. After colds, back pain accounts for the most lost work days in adults under age 45

If you are using a ladder at home or at work, make sure it is the proper height for the job at hand. Place it on a firm, even surface. If a job involves working on scaffolding, check that it is in good repair, with no missing bolts. Don't use a ladder on scaffolding: Get another tier of scaffold erected.

USING MACHINERY

Any machine with moving parts is a potential hazard: Limb entanglement can result in amputation or death. A life-threatening injury can happen in less than a second. Those most at risk include factory and farm workers. When using machines with moving parts, do the following:

- Wear all recommended protective clothing and headgear.
- Be sure maintenance guidelines have been adhered to.
- Watch out for loose clothing: Shoelaces, for example, can easily be caught in machinery.
- Adhere to advised safety distances, in particular making sure that children are not in danger (children

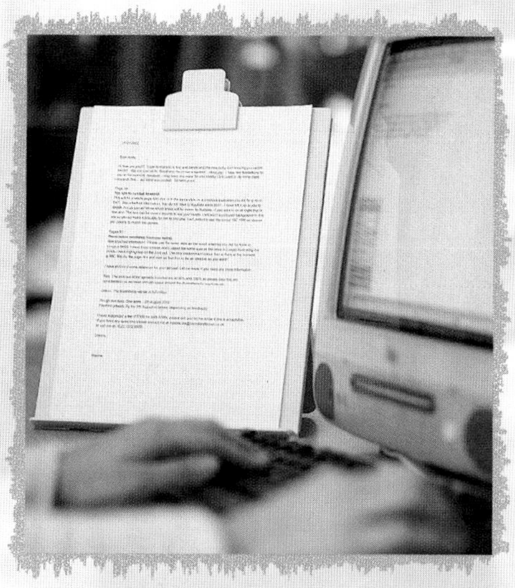

VERTICAL COPYING
Typing from copy is less stressful for the neck, shoulders, and back if the source material is placed vertically next to the screen so that your neck is not constantly bent in order to read it.

BAD VIBRATIONS
Power tools, road drills, and machinery that vibrates during use are especially hazardous for the spine. Wear well-insulated footwear and follow all safety directives.

LACE UP TIGHT
Anything dangling or hanging is an obvious hazard around machinery. Lace shoes tightly, avoiding long ends; do not wear dangly earrings, necklaces, or scarves; and fasten belts firmly.

on farms are one of the groups most at risk of being injured by a machine).

- If there is a problem, stop the machine and disconnect the power before anyone approaches to investigate.

HEALTH AND SAFETY WITH COMPUTERS

Computers are often wrongly blamed for many health problems. Certainly, when used inappropriately, they can cause pain and stiffness in the fingers, wrist, neck, and shoulders. However, there is a lot users can do to prevent such problems.

Good working practices are extremely important. A well-designed work station, where a computer user does not have to stretch unnecessarily, can help prevent injury. Some problems relating to computer work are actually caused by stress rather than the computer, so if you are finding that you are being asked to do too much, discuss this with your manager.

Prolonged use of a keyboard, and more especially a mouse, can cause pain in the fingers, wrist, and shoulder. Take frequent breaks: If your job involves a great deal of mouse use, try swapping hands, using one hand then the other for a little while (but adjust your work station accordingly). Don't keep your hand on the mouse permanently: Rest it frequently, allowing it to hang loose at your side.

Sit upright, without rounding your shoulders. Adjust your chair and screen so that they are comfortable for you. Your forearms should be approximately horizontal and your eyes level with the top of the screen. Your legs should be unimpeded, so make sure there are no obstacles under your desk and avoid excess pressure from your chair on the back of your knees. Rest your feet flat on the floor: If they don't reach comfortably, ask for a footrest.

Experiment with different mouse, keyboard, and screen arrangements to find the one that works best for you. A document holder may help you avoid awkward head, neck, and eye movements. Try to keep your wrists almost straight when you are typing and use as light a touch on the keys as possible. If you are writing or typing from copy, it pays to learn how to touch type: Good keyboard technique can avoid a great deal of distress.

Beds and back and neck health

Getting enough restful sleep is enormously beneficial to the whole body. How well you sleep depends on many factors, but among the most important are the quality of your bed, mattress, and pillows.

Three major elements combine to produce a good bed: the support it affords your body, how comfortable it is, and its durability. When it comes to support, firm isn't necessarily best. A bed that is properly supportive should provide equal resistance to all parts of your body. A bed that is too firm may only support those areas that are the heaviest, such as your shoulders and hips. A mattress that is too soft, by contrast, won't keep your spine in alignment, leading to backache and fatigue. If you lie on a mattress that is too firm, you will feel pressure on certain areas of the body; if your body sags into the mattress, it is too soft.

Comfort is subjective: What feels comfortable to one person may not to another. The key factor in determining comfort is the materials used in the bed's construction. A bed's durability also depends on the construction: How efficient the springs are in absorbing your movement on the mattress.

CHOOSING A PILLOW

In order to avoid neck pain, your head should stay in alignment with your spine while you sleep. Sleeping on your stomach is bad news for the neck: It can cause the bones to "lock," resulting in pain in the morning. For most people, only one pillow is necessary, and it should be tucked into the neck to offer the best support. It is worth experimenting to find a pillow that keeps your head neither too high nor too low. Specially designed pillows with a raised support for the neck keep the spine in alignment. These are helpful for people with persistent neckache.

A V-shaped pillow supports the back and neck of someone sitting in bed or in an armchair but is less suitable for sleeping. If you read in bed, always make sure your neck is supported by one or more pillows.

BUYING A BED

When choosing a new bed, do the following:

- Wear loose, comfortable clothing and take off your shoes. Lie on the bed (with your partner if you will be sharing the bed) and check the pressure: Your back and sides should not feel any "hard" areas.
- Buy a base and mattress together: Bed components are designed to work together. If you need to replace a mattress, check that it is suitable for the base you have. Never place a board between the base and mattress.
- Buy the biggest and best you can afford: A good bed should give 10 to 12 years wear, which amounts to a few cents a night.
- Never allow children to jump on a bed.

Pillow sense
Sleep either on your back or your side. This allows your head to stay in line with your spine; special pillows can make this easier. Your head should not be raised too high on a pile of pillows, but neither should it sink into the mattress.

Buying shoes

Poorly fitting shoes can cause pain and discomfort and may lead to long-term damage to the bone structure of the foot. Buying well-fitting shoes for everyday and sports wear will bring lasting benefits in comfort and foot health.

Only a tiny percentage of children are born with a foot deformity, but estimates suggest that by the age of five, more than 20 percent have pain or deformity, largely through ill-fitting shoes.

Shoes should match the shape of your feet, being neither too narrow, too short, nor too big. Feet may swell toward the end of the day, and for this reason it is often best to go shoe shopping late in the day. Feet are also larger when they are bearing your weight, so have them measured while you are standing. Most of us have feet of slightly different sizes, so always go for the larger size and use an insole in the shoe for your smaller foot, if necessary.

The two most important areas of the shoe are across the toes and the heel. Toes should fit into the shoe without pinching or bunching, and the heel should not slide around. Low heels—no higher than 1½ inches and with a broad base—are kinder to feet. Straps or laces will ensure that the foot doesn't slip around in the shoe as you walk. Always lace shoes correctly: Unlaced or poorly laced shoes encourage toe curling to hold shoes in place, and this stresses the bones of the toes.

Shoes with high heels and pointed toes can damage the toes, ankles, calves, and back; they have also been implicated in headaches and neck pain. Toe deformity that can only be corrected by surgery may result. The higher the heel, the greater the pressure on the ball of the foot, which can cause long-term problems.

It should never be necessary to "break shoes in": Shoes should feel comfortable from the moment you put them on. If they don't feel comfortable, don't buy them.

SHOES FOR SPORTS

There is a vast array of sports shoes on the market, many of which are expensive. It is not worth compromising bone and muscle health for the sake of a few dollars, so it is worth investigating exactly what you are getting: "Designer" does not necessarily equal better for you.

Cross-trainers are, as the name implies, intended for several sports. Their detractors say that they don't offer as much support as sports-specific shoes, which may be true, and anyone who only jogs or plays squash, for example, is probably better off buying the appropriate shoes for their chosen activity. Many people, however, try different activities at different times, and cross-trainers provide enough stability, cushioning, and comfort for most popular sports. They are also

0-18 YEARS

Shoes for children

The bones of the foot are not fully "set" until a child is about 18, so poorly fitting shoes in childhood really can make a difference to adult foot health. Children may not be able to tell you if their shoes are hurting, so accidental damage can result. Always buy children's shoes from a reputable store and get the child's feet measured for both length and width. Flat shoes are best for children, although a pair that is only worn occasionally could have a small heel. Slip-ons should be avoided for children: Choose a style that is secured by laces, buckles, or velcro. Breathable materials, such as leather or canvas, allow sweat to evaporate, preventing infections such as athlete's foot. There should be a child's thumb width of growing room between the longest toe (not necessarily the big toe) and the front of the shoe. Finally, bear in mind that children's feet can grow at different rates from child to child and from one year to the next, so one pair of shoes may last a shorter time than another.

BEING SHOE WISE

These are the factors to consider when assessing whether a pair of shoes will be comfortable and problem-free.

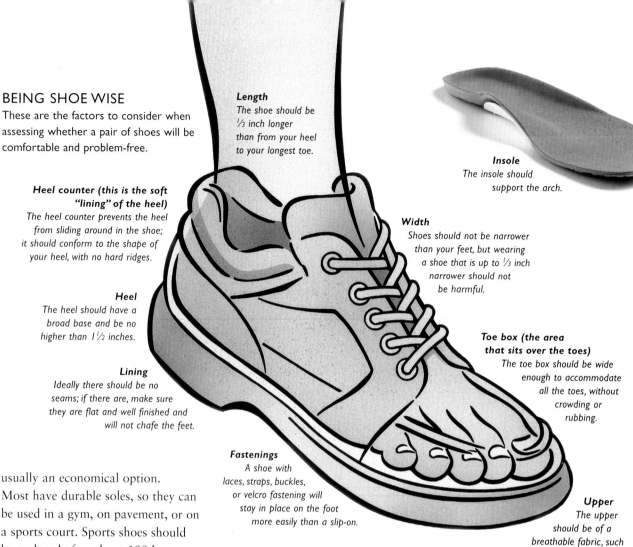

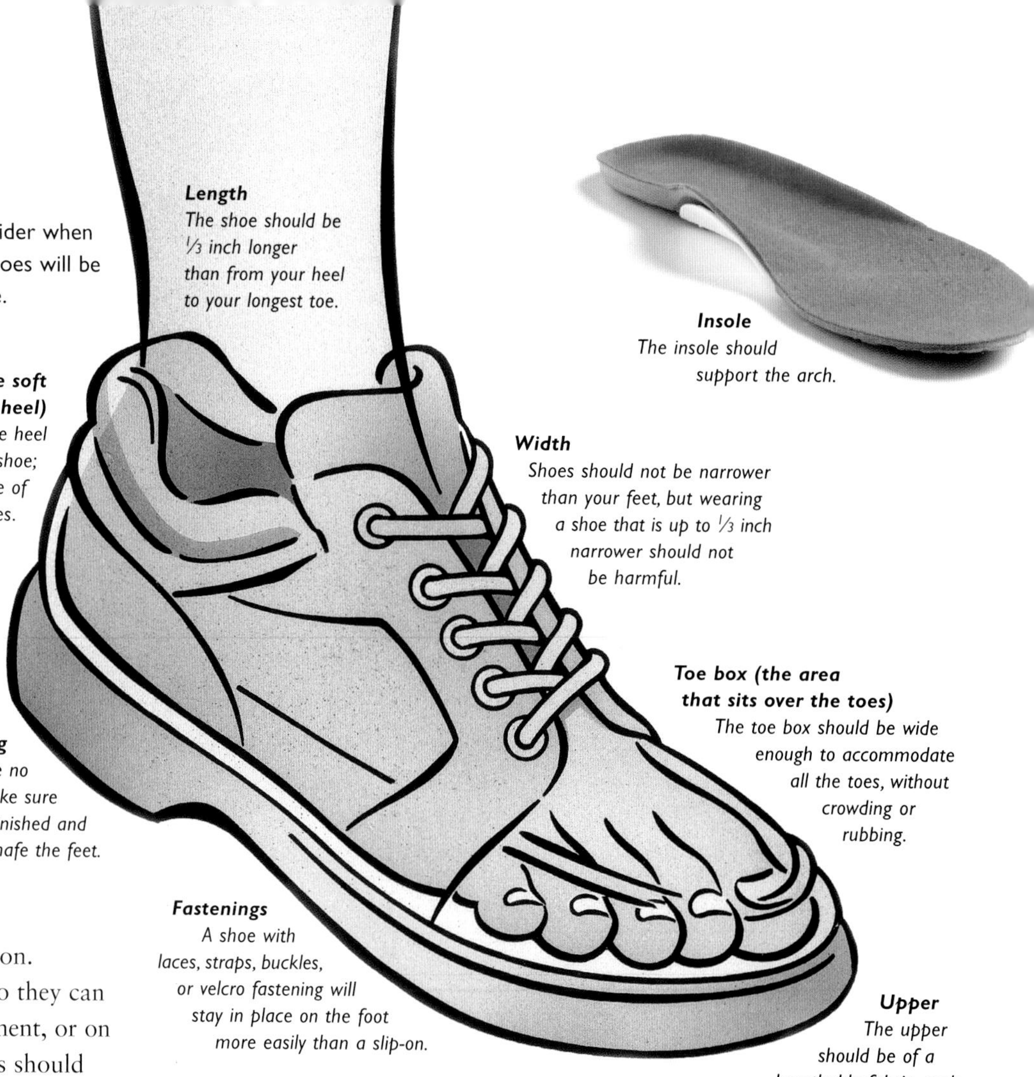
Length
The shoe should be ⅓ inch longer than from your heel to your longest toe.
Heel counter (this is the soft "lining" of the heel)
The heel counter prevents the heel from sliding around in the shoe; it should conform to the shape of your heel, with no hard ridges.
Heel
The heel should have a broad base and be no higher than 1½ inches.
Lining
Ideally there should be no seams; if there are, make sure they are flat and well finished and will not chafe the feet.
Fastenings
A shoe with laces, straps, buckles, or velcro fastening will stay in place on the foot more easily than a slip-on.
Width
Shoes should not be narrower than your feet, but wearing a shoe that is up to ⅓ inch narrower should not be harmful.
Toe box (the area that sits over the toes)
The toe box should be wide enough to accommodate all the toes, without crowding or rubbing.
Upper
The upper should be of a breathable fabric, such as leather or canvas.
Insole
The insole should support the arch.

usually an economical option. Most have durable soles, so they can be used in a gym, on pavement, or on a sports court. Sports shoes should be replaced after about 100 hours of wear—that is, once a year for someone who has two hour-long exercise sessions a week.

Running shoes need plenty of shock-absorbing cushioning in the sole to protect feet from hard ground; they also need to be lightweight and offer good support from heel to toe. Trauma is common in runners, so anyone who runs often should replace shoes regularly.

High-sided hiking boots are the best choice for anyone who walks seriously in open country. They are designed to give good traction on uneven surfaces and to protect the heels and ankles from twists on uneven terrain. They are less essential for people who walk in cities, where the pavements are more even: Cross-trainers or walking shoes are adequate.

Shoes for tennis or squash should offer good side-to-side support to avoid injury to the ankles caused by the sideways movements these sports require. Similarly, basketball shoes should offer good ankle support.

Canvas shoes don't give much support to the arches, so they are not a good idea for high-impact sports such as tennis or for road running. Sports shoe technology has moved on since the first canvas sports shoes were produced. There are now many "trendy" sports shoes on the market, but always choose the most supportive for your training needs.

If you play a sport that needs spikes in your soles—football, for example—opt for molded plastic rather than metal spikes. These are safer if you slide into an opponent but offer just as much grip.

Overpronation

Up to an astonishing 85 percent of the population are overpronators, that is, when the foot hits the ground while walking or running, it rolls inward, increasing the strain on the muscles, tendons, and joints of the foot. A similiar problem is the lowering of the arch so that the sole lies flat on the ground. This is referred to as flat feet. Although both conditions result in many injuries each year, it is possible to buy specially designed shoes and insoles that can help. Shoes that provide motion control and stability are ideal for overpronators, and supportive insoles can relieve the problems caused by flat feet. There are specially designed insoles for flexible or rigid flat feet.

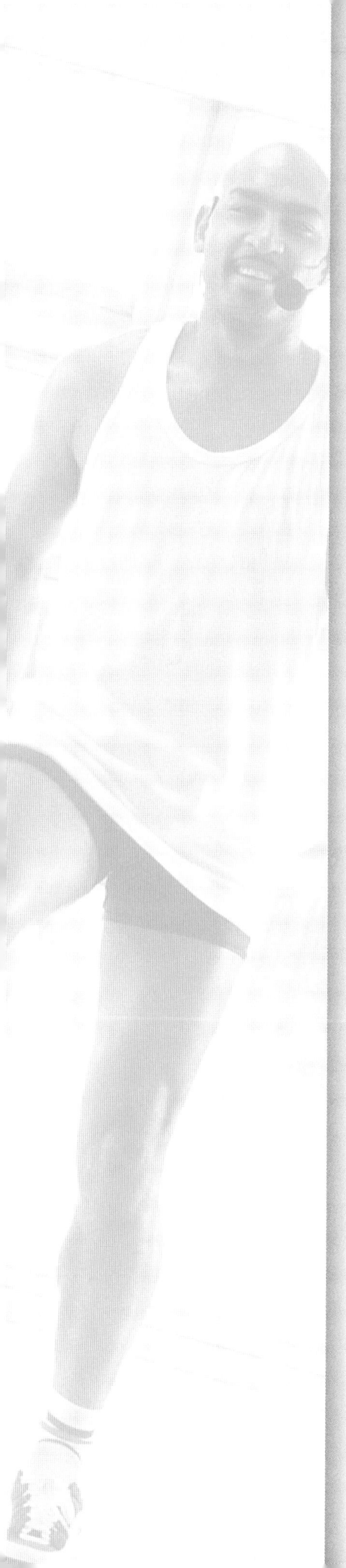

EXERCISE FOR BONE AND MUSCLE HEALTH

Exercise builds and firms muscles and improves your general posture. It also builds and preserves bone health, helping to delay the onset of osteoporosis. The key is to choose activities that you enjoy, because this makes it much easier to do them regularly.

86

Getting regular weight-bearing exercise will promote strong bones and muscles and help preserve bone health in later life.

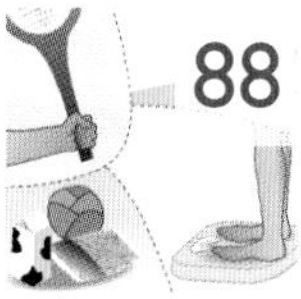
88

Osteoporosis can be a debilitating condition, but there's a lot that can be done to prevent or delay its onset and to minimize its impact.

90

An exercise session does not have to take place in a gym. A workout at home, using weights or everyday items, also benefits bones and muscles.

92

A bone or muscle problem does not have to be a barrier to getting exercise, but you may need to be more careful in your choice of activities.

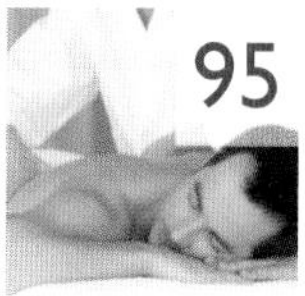
95

Massage relaxes tired muscles and, together with RICE (rest, ice, compression, and elevation), speeds healing after a minor injury.

Choosing the right exercise

Exercise has enormous health benefits for the whole body and is the best way to build and maintain strong bones, to keep joints mobile, and to increase muscle mass. Choosing the right type of exercise is important.

Exercise keeps the muscles and bones in good working order and keeps the joints mobile and flexible, too. The benefits for your musculoskeletal system, and in turn, your whole body, are huge. Generally speaking, exercise increases muscle tone and flexibility, improves overall fitness levels, slows down the aging process, and improves your overall sense of well being.

Physical activity increases your intake of oxygen and stimulates the flow of blood around your body. The heart beats faster and more strongly, speeding up circulation of the blood. This has a direct impact on your bones and muscles by providing them with a powerful dose of the nutrients they need. The increased blood flow to the bones helps them retain their density, which protects them against injury.

A 1992 study showed that regular walking can reduce osteoarthritic knee pain by 27 percent.

The rate and depth of your breathing also increase when you exercise. This improves the oxygenation of the blood, providing your muscles with the oxygen that they need to function well. Exercise strengthens the muscles, as well as the ligaments that attach them to the bones. This in turn improves the strength and mobility of the joints. Even damaged joints can have their mobility restored, and certain types of back pain can be eased using specifically designed exercises.

HOW MUCH EXERCISE IS ENOUGH?

There is no specific formula for frequency and intensity to prevent bone and muscle problems. To benefit from exercise in general, however:

- Your chosen type of exercise needs to be repeated several times a week in order to benefit from it. Ideally you should be exercising 3 to 5 times per week for 30 to 60 minutes each session.
- Exercise intensity needs to be fairly vigorous; breathing should be increased and you need to feel the muscles working.
- Rest should be taken between work-outs to allow the muscle fibers time to recover. Variety of exercise is also recommended for the noncompetitor.
- Increased frequency or high-intensity exercise can lead to injuries and often makes athletes liable to infection. Very intense exercise is only recommended for the professional and semiprofessional athlete.

WHAT KIND OF EXERCISE?

Exercise is vital for developing strong bones and muscles. However, your physical strength, heart health, and bone condition must be adequate for the level of exercise. Your doctor or a fitness instructor can assess what is safe and effective for you.

WEIGHT-BEARING EXERCISE

Weight-bearing exercise is activity in which weight is placed on the skeletal frame and then pushed away from it. A gravitational and muscular pull, for example, is put on the skeleton as it is moved up and down by the muscles. As you run, jump, or dance, the force of movement is transferred through all the supporting bones of the body. The impact of this stresses the bones, and the force of the muscles pulling on them (to produce movement) stresses them further, demanding and creating extra bone strength.

Weight-bearing exercise includes walking, running, jogging, step and aerobic classes, and dancing. Most of these types of exercise are aerobic, so they will also increase the performance of the heart and lungs and reduce body fat.

STRENGTH TRAINING

Strength training (see pages 90–91) involves activity that improves muscle mass through muscle resistance. Free weights and weight machines are popular ways to build strength. Strength training can be performed as little as twice a week and needs not involve special equipment other than simple weights or elastic exercise bands.

IMPACT WORK

Impact work is activity that involves the striking or hitting of something. When you jump and land or when you strike a tennis ball, for example, you are causing increased impact on the skeleton. This kind of exercise holds a greater risk of injury than weight-bearing exercise, but it can be invigorating and extremely beneficial in stressing the bones and making the muscles stronger. X-rays of a tennis player would show increased bone mass in the racquet arm. Examples of impact exercise include tennis, basketball, volleyball, skipping, and gymnastics.

FLEXIBILITY WORK

Much of our day is spent sitting or bending. This tightens our muscles and puts a lot of stress on the spine. Exercises such as yoga, pilates, and tai chi gently stretch the muscles and develop spinal flexibility. At home, you can improve flexibility by stretching the muscles—to a point of tension, not pain—and holding each stretch for 15 to 30 seconds twice daily. You can even stretch at your desk. Avoid forward bending exercises to prevent extra stress on the bones and discs of the spine.

HOW TO START

Follow a gradual approach to exercise to get the most benefits with the fewest risks. If you have not been exercising, start at a slow pace and as you become more fit, gradually increase the amount of time and the pace of your activity.

Choose activities that you enjoy and that fit your personality. For example, if you like team sports or group activities, choose something like aerobics. If you prefer individual activities, choose swimming or walking, for example. Also, plan your activities for a time of day that suits you. If you are a morning person, exercise before you begin the rest of your day's activities. If you have more energy in the evening, plan activities that can be done at the end of the day. You will be more likely to stick to exercise if it is convenient and enjoyable.

WARM UP AND COOL DOWN

Proper preparation for any sports activity will ease or prevent cramps and reduce the risk of muscle strain.

RHYTHMIC MOVEMENTS
Heating up the muscles with gentle activity at the start of an exercise session will ensure that the blood vessels are dilated and stimulate blood circulation. This will maximize the amount of activity that the muscles can endure before suffering from cramps or other injuries.

STRETCHES
Proper stretching is essential: Slow muscle stretching will help avoid muscle strains by increasing flexibility. Aim to stretch at least your back, chest, shoulders, arms, and legs. If you do get a cramp, press on the muscle, then stretch out the affected part and rub. Repeating the stretches at the end of an exercise session will help you avoid muscle soreness.

HEALING MASSAGE
Gentle massage and warm baths may also help reestablish the circulation after cramps, but they do not make any drastic attempt to manipulate the affected part. Gentle rubbing and kneading with the hands encourages blood flow to stiff muscles.

The benefits of exercise

Exercise is one of the key factors in building and maintaining strong bones and muscles and thus aiding in the prevention of a range of conditions in later life, including osteoarthritis and osteoporosis.

BUILDING BONE MASS

Regular exercise has beneficial effects on bone health at all ages: Early in life it is known to promote higher peak bone mass, during midlife it can help to prevent bone loss, and later in life regular exercise can help to slow down the loss of bone.

Bones, like muscles, are living tissue: They get broken down and rebuilt in an ongoing process called remodeling, which balances formation and resorption. If the body is not exercised enough, the bones lose calcium, and become weaker. It has been noted that patients on enforced bed rest can lose significant bone mass.

IT'S NOT TRUE!

"Osteoporosis only affects women"

Although women tend to be at a higher risk, men also suffer from osteoporosis. In fact, in Singapore men get osteoporosis much more frequently than women, and male osteoporosis is on the rise in the West: In the United States 25 to 30 percent of hip fractures occur in men. The disease can take hold when men's testosterone levels are low—which can be caused by medications such as corticosteroids, by health conditions such as liver disease, or by smoking.

Astronauts have been found to lose bone density while in space at a rate of 1.5 percent a month. This is because there is no gravitational resistance to keep the bones and muscles stimulated. NASA experts discovered one of the ways to combat this was weight training.

It seems that the action of the muscles pulling on the bones—when you lift weights, for example—stimulates small electrical changes. This is the body's way of telling it to be ready to handle that weight of impact again. This stimulation promotes new bone growth. Exercise also seems to affect the hormonal control of the osteoclasts and osteoblasts which break down and re-form bone. Finally, exercise means increased blood supply to the bones, which again aids bone mass.

Exercise has been shown to prevent or reverse at least 1 percent of bone loss per year.

The type of exercise you choose is important. Most experts believe that two types of exercise provide the most bone benefits: weight-bearing exercise and strength training (or resistance) exercise.

Weight-bearing exercises are those in which bones and muscles work against gravity. These include any exercise in which the feet and legs bear the body weight, such as jogging, walking or hiking, climbing stairs, dancing, and racquet sports. Strength training involves the use of weights during exercise. The use of free weights appears to stimulate bone growth in the hips. To increase bone density in the spine, use either free weights or exercise machines. Ask an exercise professional to design a well-balanced strength program for you involving all the major muscle groups, particularly those in the hips, wrists, and spine—these are common osteoporosis fracture sites.

To improve bone health you must challenge your muscles with 8 to 10 repetitions of a movement 2 to 3 times per week. It is important to slowly increase the weight as you become stronger rather than sailing through a workout that has become easy: Exercise only works if you keep at it. If you stop exercising, the benefits start to be lost after two weeks and disappear after a few months.

Less effective activities

Bicycle riding and swimming, although effective aerobic exercises, are not weight-bearing exercises. They should only be used alongside weight-bearing or strength activities in a bone-health program. They do, of course, provide many other health benefits, including flexibility and cardiovascular fitness.

INCREASING MUSCLE STRENGTH

Strong muscles give support and balance to the body. If a muscle is unused, it atrophies and wastes away. In contrast, muscle that is exercised becomes thicker and stronger. Exercising with weights develops muscles and ligaments, increasing their strength and endurance. It also helps improve posture.

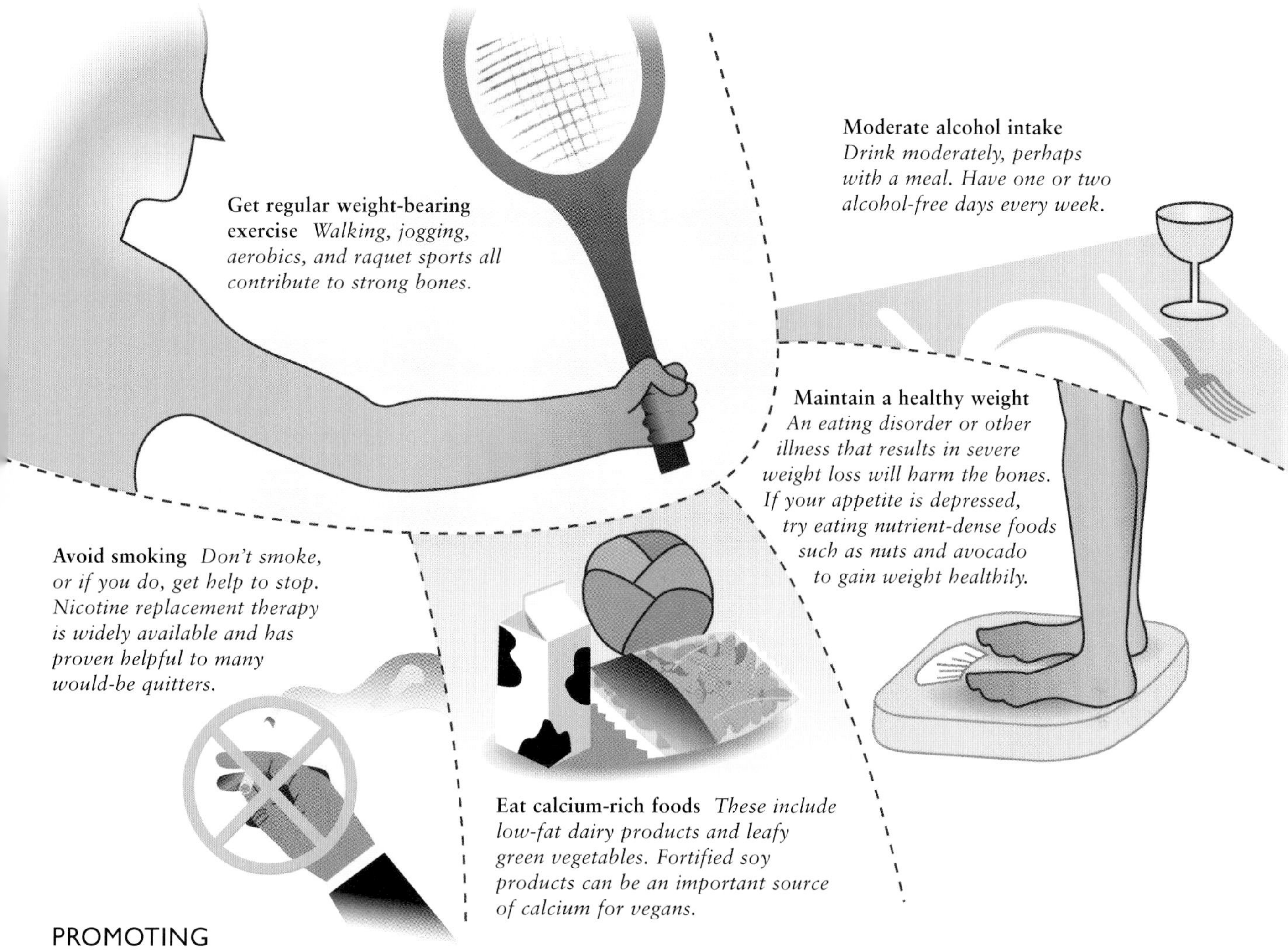
Get regular weight-bearing exercise *Walking, jogging, aerobics, and raquet sports all contribute to strong bones.*
Moderate alcohol intake *Drink moderately, perhaps with a meal. Have one or two alcohol-free days every week.*
Maintain a healthy weight *An eating disorder or other illness that results in severe weight loss will harm the bones. If your appetite is depressed, try eating nutrient-dense foods such as nuts and avocado to gain weight healthily.*
Avoid smoking *Don't smoke, or if you do, get help to stop. Nicotine replacement therapy is widely available and has proven helpful to many would-be quitters.*
Eat calcium-rich foods *These include low-fat dairy products and leafy green vegetables. Fortified soy products can be an important source of calcium for vegans.*

PROMOTING HEALTHY BONES

Using exercise to build strong bones is half the story, but lifestyle factors can reverse all your good efforts. Exercise as part of a healthy lifestyle is best for your bones.

Mobility and balance

Strong muscles play an important role in reducing the risk of falls in elderly people by improving overall strength, balance, and mobility. It decreases the propensity to fall and improves the ability to absorb impact. Exercise programs for the elderly can reduce the risk of falling by 10 percent, and exercise that helps to improve balance—such as dancing—can reduce the risk of falling by up to 20 percent.

PREVENTING BACK PAIN

Back muscles do not get adequate exercise from normal daily activities. They also show a natural tendency to weaken with age unless they are specifically exercised. Exercises that stretch and strengthen the muscles of the spine can help prevent back problems. If back and abdominal muscles are strong, it becomes easy to maintain good posture and keep the spine in its correct position.

Almost all back pain originates in the muscles (rather than bone or other structures), so exercise to prevent back pain should focus on muscular fitness. Yoga is ideal: The postures gently stretch and strengthen muscles, and, in addition, the spine becomes more flexible, giving the body a greater range of motion. It is an excellent form of exercise for preventing backaches.

REHABILITATION

Specific exercises can be used after an injury to reduce swelling, prevent stiffness, and restore normal pain-free movement. When the acute pain and swelling have diminished, a patient may be instructed to do a series of exercises several times a day. The next goal is to increase strength and regain flexibility, with increased exercises as function improves. The final goal is to return to full daily activities, including sports (see Exercising after injury, page 92).

A weights workout

Regular exercise with weights is an excellent form of resistance training. It develops bone and muscle strength and should form an integral part of any exercise program. Try to fit these simple exercises into your weekly schedule.

Lifting weights helps build and preserve bone density and increases muscle tone and flexibility. It is a key part of any exercise routine designed to lessen the risks of osteoporosis in later life.

Resistance exercises work the muscles, which pull on the bones via the tendons and ligaments; all elements are strengthened in the process. Lifting weights will also improve your balance, coordination, and confidence, all of which should help prevent falls.

You can perform this routine every second or every third day to build strength in your muscles and bones, but give your body time to recuperate. Use a set of dumbbells if you have them (these are widely available from sports stores in a range of weights: start low, say two pounds). If you don't have weights, cans of soup or beans can provide a weight when you are starting out. In all of these exercises, lift and lower the weight slowly, maximizing muscle strength while minimizing the risk of injury. If exercise starts to become painful, stop and seek advice from a trainer at the local health club.

Perform the routine once to begin and then rest for a day before repeating it. As you get stronger, you can perform the whole routine two or three times in succession for a more challenging workout. When this becomes easy, start to think about increasing the weights you are holding. The "right" weight is one that makes the last couple of moves a bit of a challenge: If it all seems too easy, you need to increase the weight.

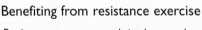

Benefiting from resistance exercise

Resistance work means working the muscle against a force that is resisting. As you contract the biceps muscle in the front of your arm to lift a weight, for example, and then lower the arm slowly, you are resisting the force of gravity on the way down. This kind of exercise contracts the muscles, which pull on the ligaments, tendons, and bones, sending out all-important electrical impulses. It also increases strength in the muscles, thereby improving posture and coordination. One of the best ways to provide this kind of exercise is by lifting weights. You can also use resistance bands, weighted balls, and your own body weight to provide resistance. Studies have suggested that regular resistance exercise enables older people to keep up to 80 percent of their young adult strength. Weightlifters have also been shown to maintain their fast muscle twitch fibers—used in fast movement such as sprinting—which most of us tend to lose as we get older.

1 Lunge and pull

Step one leg in front of the other, holding your hands in front of you at hip height, just touching with a weight in each hand. Bend the knees so that your hips drop directly toward the floor; your back knee will be just above the floor. At the same time, pull your arms up with elbows out to the sides and the weights coming up to chest height. Lower your arms slowly as you push your legs straight. Repeat this move 10 to 12 times, on each leg.

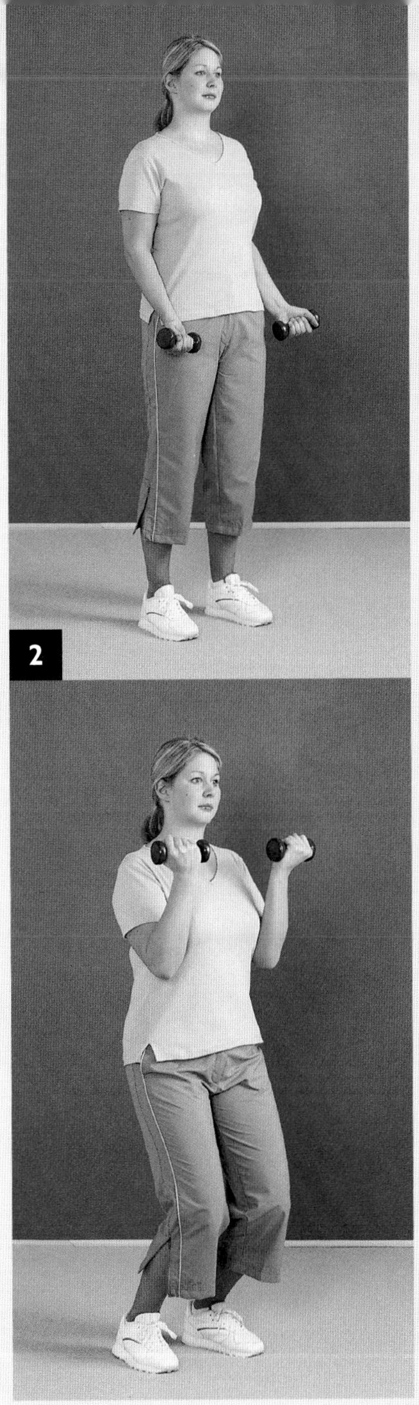
2

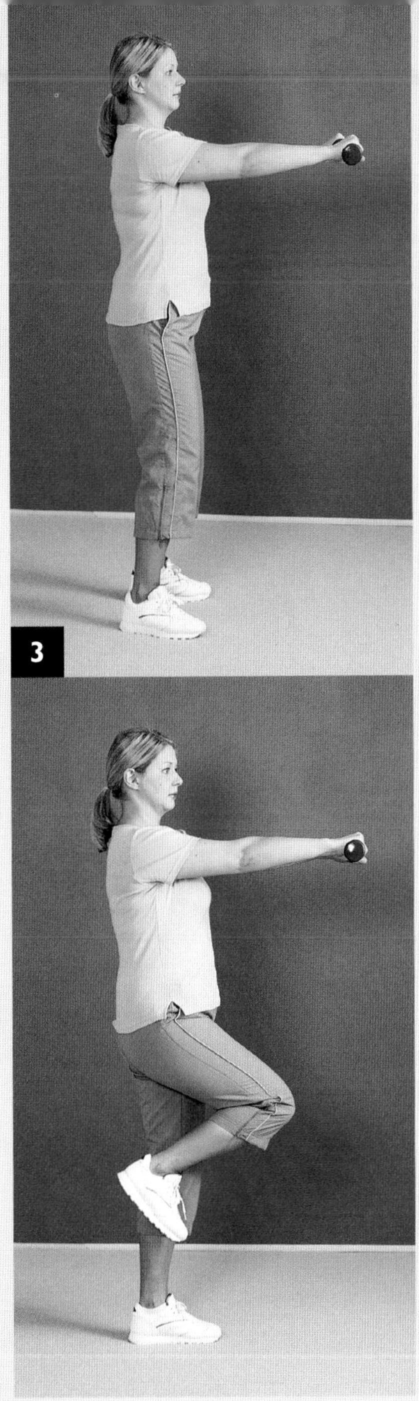
3

4

2 Biceps curl and squat

Stand with correct posture (see page 72) with a weight in each hand, palms facing the front. Contract the muscles in the front of your arms (biceps) and bend the arms up. As you do this, sink your bottom behind you, bending the knees: Keep your back straight. Now press through your legs to straighten and at the same time lower your arms slowly. Repeat this movement 10 to 12 times.

3 One man mow

Stand evenly on both feet, in correct posture, with a weight in each hand and palms facing downward. Lift both weights straight out in front of you to shoulder height. As you do this, transfer your weight onto one foot and lift the other foot up to the side of the knee. Lower the weights as you replace the foot, transferring your weight back onto both feet; this should be done gently without stamping. Repeat the move 10 to 12 times, alternating legs.

4 Plié push

Stand with your feet in a wide position and slightly turned out. Take the weights up to your shoulders, with your elbows tucked into the sides. Bend your knees, pressing them out over the direction of your feet as you lower your bottom. As you bend your knees, press your arms up into the air with the weights pushed straight toward the ceiling. Bend your arms and straighten your legs simultaneously in slow, smooth movements. Repeat the move 10 to 12 times.

Exercising with bone or muscle problems

A bone or muscle problem does not have to be a barrier to exercise. After an injury or when the body is otherwise compromised, perhaps by arthritis or osteoporosis, exercise still has an important role to play.

EXERCISING AFTER INJURY

After an injury—following a period of treatment and rest—many people find that although the injury appears to have healed, they are unable to return to exercise without aggavating the injury. This injury–rest–reinjury cycle is common, and the key to breaking it is to follow the principles of "reconditioning."

Reconditioning involves safely working toward former fitness levels by carefully controlling exercise intensity and using pain to gauge acceptable levels of exercise.

It may be helpful to think of pain in two ways. Type I pain is pain felt during exercise. Type II pain is the residual pain felt about 90 minutes after exercise. Following an injury, use type II pain to monitor the level of exercise. After gentle exercise, if there is no type II pain, or if there is only mild pain that is relieved by an hour of light activity, you can slightly increase your exercise level. If type II pain is intense, you must rest for 1 to 3 days, then resume the reconditioning at a lower intensity.

Start with exercises that involve static, pain-free stretching. When flexibility has improved, use resistance exercise (see page 90) to increase muscle strength. When you are sufficiently strong, gradually increase the intensity of your session.

WORKING OUT

Women reluctant to start hormone replacement therapy (HRT) or to take medication to limit bone loss might consider working with a personal trainer. Bone loss still occurs, but most women can avoid weight gain, maintain good general health, and continue to play sports through having an exercise program. Although falls may still happen, they are less likely to lead to fractures.

EARLY-STAGE OSTEOPOROSIS

In the early stages of this condition, weight-bearing exercises such as walking, dancing, and stair climbing can be used to slow down bone loss.

Resistance work—using weights—can still be done. The weight you use should be such that, after 10 to 12 repetitions, you are unable to lift more because the muscles are fatigued. This weight is called your repetition-maximum, and it will be different for everyone. Lifting this kind of weight will ensure that your muscles are working hard enough to gain some benefit without being asked to work so hard that you sustain an injury.

Other types of exercise that are beneficial include aerobics and step aerobics. These kinds of exercise have been designed for weight loss and muscular toning. They are also aimed at keeping the participant going for a minimum of 20 minutes (to enable greater stores of fat to be burned) and therefore have a rhythmic, regular momentum.

LATER STAGE OSTEOPOROSIS

In the later stages of this condition, body weight alone can be enough to cause a fracture—for example, of the spine. Water aerobics and swimming are good in these cases, because the water provides a good resistance while supporting the whole body weight. Exercises for strengthening the upper and lower back can be done in the water. Work on posture, balance, and coordination becomes more important once the disease has progressed. In the worst cases, only bed exercise may be recommended.

Stiffness in the initial stages of an exercise program is normal, but if the pain is too uncomfortable, ask your doctor for advice.

Osteoporosis causes more than 300,000 hip fractures, 250,000 wrist fractures and 700,000 spinal fractures a year in the United States.

Exercises to avoid

- Forward flexion—that is leaning the upper body forward—which can cause vertebral fractures.
- Stooping and rotating the shoulders, which also puts you at risk of fractures.
- Rowing machines cause deep forward bending, which can lead to fractures.
- Skiing leaves you open to too many fall possibilities.
- Running and tennis are too jarring.
- Bowling and golf involve too much percussive twisting and rotation.

Move and breathe

Water aerobics gives your heart a workout and also improves muscle tone. There is less likelihood of sustaining an injury than with aerobics, because the water cushions the legs and spine, with no force transmitted, as through a gym floor.

AEROBICS

Join an aerobics class that is appropriate for your age and current state of health. An ideal aerobic program for the older woman, for example, might include gentle warm-up and mobilization exercises to increase circulation and range of motion. Choreographed, low-impact moves will build heat and get the breathing going. If a class does not appeal to you, you could try doing an exercise routine at home.

Sample home aerobic routine

Put on some medium-paced music and do each of these for 3 to 4 minutes, with the whole routine lasting 20 to 25 minutes.

- March up and down on the spot. Swing your arms and lift your knees up as high as you can. Keep the movement smooth and step on your toes, rolling onto your heels as you step.
- Start to step from side to side in time to the music, and clap your hands gently in rhythm.
- Begin to take four or five paces forward, bending your knees in time to the music and swinging your arms as you go. Then come back, still facing the same direction.
- Now march on the spot again, pulling up your abdominal muscles to support the torso.
- Kick one leg out in front of you and then replace the weight back on that leg. Kick again and punch with your arm as well. Kick and punch several times then change legs. Make sure when you kick that you are balanced properly on the other leg. Keep your abdominal muscles tight to help you balance.
- Now try kicking your leg behind you as well.

During the exercise, you should feel your muscles working and your heart pumping, but you should always feel in control of the movements you are making.

People who have undertaken programs similar to this have shown improvements on tests of balance, agility, and coordination, thereby reducing their risk of injury from falling. This kind of program also provides an overall workout that will keep the heart and lungs in shape and help in the fight against weight gain. People who get regular exercise have a greater body awareness and confidence. These factors also help reduce the chance of the uncoordinated falls that often lead to breakages. Any routine should involve the following:

- Strengthening exercises for the legs, hips, and buttocks to improve the functionality of the entire area so that this, and not the spine, is used when lifting.
- A range of exercises in both seated and standing positions.
- Abdominal and lower spine exercises to strengthen and stabilize the torso.
- Dynamic balance exercises to improve balance and reduce the risk of falling.

Swimming and the bones

TALKING POINT

Swimming is very beneficial in terms of general health and fitness. It improves flexibility and cardiovascular fitness, provides social opportunities, and can be very good for older people. It has been suggested in many studies, however, that swimming does not place any significant stress on the bones. In fact, some studies, in comparing swimmers with other athletes, have shown that they have the same bone density as someone who gets no exercise at all. It seems even the pull of the muscles and the increase in muscle mass does little to increase actual bone density without the added force of gravity.

TYPES OF EXERCISE FOR OSTEOARTHRITIS

If you are in any doubt about the suitability of a sport or exercise routine, consult a qualified physical therapist or a personal trainer at your local health club. Physical therapy will be recommended after surgery for a bone or joint problem (see page 126).

- Exercise that increases flexibility, such as stretching, will help mobilize inactive joints. Good choices include yoga, tai chi, and pilates.
- Although weight-bearing exercise is good for bones, it can stress already compromised joints and so may not be appropriate soon after surgery or if arthritis is a problem. Swimming, water aerobics, and gentle cycling are less stressful.
- Everyday activities—walking, stair climbing, gardening, housework—also count as exercise and keep the joints moving.

Flowing moves
Low-impact exercise such as tai chi helps improve balance, agility, and coordination. Many health clubs now offer classes in tai chi.

Water workout
Swimming, although not weight-bearing, will provide a good cardiovascular workout and is low risk in terms of falls. Webbed gloves allow you to use the water as a resistance so that your arm muscles get more of a workout.

Walk wise
Hiking is great exercise in the fresh air, but choose your terrain carefully if you have arthritis or have suffered from a fall. Uneven ground can be a hazard.

Healing damaged tissues

Rest, Ice, Compression, and Elevation (RICE) is an effective treatment for minor injuries such as a sprain, muscle pull, or impact injury. Learning to relax can help to prevent a recurrence, and massage may speed healing.

REST

First and foremost, you should stop doing the activity that caused the injury. This may seem obvious, but many people ignore pain in the hope that the problem will disappear if they carry on as normal. This is not the case. The purpose of a rest period following an injury is to give the body a chance to recover by itself, as well as to help it along using additional methods as described here.

Generally, it is thought that an injury such as a sprain or strain needs to be immobilized for 3 to 6 days following the injury to allow healing. You can use the level of pain as an indicator. Pain is the body's way of making us stop and take stock. As an injury heals, the level of pain subsides and that is an indication to start moving again.

However, immobilization is not recommended for long periods. If, for example, you have injured your shoulder, you might avoid activities that require overhead lifting, but you should try to keep the joint moving through its normal range of movement (as far as is possible) to avoid stiffening or loss of mobility.

ICE

Cooling down the injured area is the next stage of recommended treatment. Place a cold wrap or an ice pack wrapped in a towel on the affected area. A package of peas (or any frozen vegetable) is ideal, or you could try a plastic bag filled with crushed ice. You will probably have one or the other handy, and the pack will conform to your body shape.

Applying an ice pack lessens swelling. Although swelling initially helps to immobilize the injured part, it also restricts the flow of oxygen to the tissues and can cause cellular damage if allowed to continue. Cooling the area reduces swelling, spasms, pain, and inflammation and should be applied within 24 hours of the injury. Ice allows the injury to heal itself more quickly.

Rest:
An injured muscle needs rest to aid its recovery. Pain caused by a movement is the body's way of telling you to stop. Start moving the affected area again when the pain subsides.

Ice:
Ice helps to reduce swelling, which in itself causes more pain and slows healing. A packet of frozen vegetables wrapped in a towel is reusable and will mold itself to the injured area.

Compression:
An "Ace" bandage also helps to reduce swelling. Do not wrap the bandage too tightly: It should not cause pain, and any fingers and toes beyond the bandage should stay pink and not feel tingly.

Elevation:
Raising the injury above the level of the heart may reduce any throbbing pain. Prop up an arm or a leg. You may need to be lying down to elevate an injury to the leg.

Ice or heat?

It is important to know when to use ice and when to use heat. Generally, icing helps calm swelling. Heating an area promotes blood flow and the carrying away of waste products and is usually applicable following the icing period. Soreness or tightness will benefit from heat treatment.

ASK THE EXPERT

The length of time to use ice can vary depending on how much body fat you have and the depth of the injury. If you are slim and the injury is on the surface, 10 minutes may be sufficient. If you have a deeper layer of fat, up to 30 minutes may be appropriate. It is generally thought that 30 minutes is adequate, however, and you should not overdo it. Excessive icing can cause its own damage. Two to three daily icings should be enough. However, if the pain is severe, the affected area could be iced every one to one and a half hours.

COMPRESSION

Compressing the affected area with an elastic ("Ace") bandage or wrap can help reduce further swelling. For an acute injury, immediate compression is important. Always wrap from the largest muscle above the injury to the largest muscle below it. Do not wrap so tightly that blood flow is restricted. Fingers or toes beyond the bandage should stay pink and not feel tingly. Compression can be done during the cooling process by taping the ice pack under the bandage (but not next to the skin). This should be continued throughout the day. At night, the compression bandage should be removed and the injury elevated.

ELEVATION

Elevating the injured area helps reduce swelling and any internal bleeding. Ideally, the injured area should be elevated above the level of the heart; you may need to lie down. This reduces bleeding and helps prevent pooling of fluids by aiding the venous return. It can ease any throbbing pain. Elevation is a particularly good idea at night, when the bodily processes slow down and healing increases.

ALTERNATIVE—EDUCATION

It has been suggested that the E of RICE should also stand for education. A good understanding of what is happening as the body recovers from injury helps to avoid making things worse, and medical advice should always play a part in any treatment plan, so consult your doctor or therapist. Some consultants can recommend alternatives to painkillers, for example, and they can check that your methods are not aggravating the condition.

TOO BUSY TO RELAX?

Everyday life is busy and stressful for many of us. Few of us stop to relax on a regular basis, and this can have a detrimental effect on our bodies.

Much of the stress of everyday life gets concentrated in our muscles one way or another. Different emotional states tend to result in tension of specific areas of the body—anger tightens the chest muscles, for example, and anxiety those of the neck. The tension that results reduces the blood circulation and therefore diminishes oxygen supply while increasing the accumulation of waste products, such as carbon dioxide and lactic acid. Muscle tension can therefore lead to fatigue and so contribute to further stress.

RELAXATION ROUTINE

The following exercise encourages relaxation and the release of tension from both mind and muscles:

- Lie on your back with your arms resting by your sides and your head aligned with your spine.
- Take several breaths, each one becoming increasingly deeper, and release slowly and deeply. The final breath should be drawn right down into the stomach.
- Now focus on your neck and relax the muscles there. Push your head into the floor and then release. Use this feeling of release to relax the muscles further. Become aware of how the weight of your head is fully supported by the floor.
- Move your focus to your shoulders and repeat the process. Press your shoulders into the floor and then release. Register the fact that your shoulders are now supported by the floor and do not need to be held by your muscles. Make a conscious effort to relax the muscles in the shoulder area.
- Work through the whole body in this way, moving on from the shoulders to the ribcage, stomach, hips, legs, and feet.
- When you have reached your feet, your body should feel heavy and fully rested on the floor. Continue to breathe for a few minutes in this relaxed state.

MASSAGE

Massage can provide welcome relief from the symptoms of muscle pain. It is frequently recommended for the treatment of minor sports injuries and repetitive stress injuries.

- Massage stimulates blood circulation. This improves delivery of oxygen and nutrients to the muscles and the removal of lactic acid, which can cause stiffness.
- Massage speeds up the healing of minor sprains and strains if used after the initial inflammation has decreased.
- Massage relaxes the muscles, leading to increased range of motion. Ligaments and tendons increase in suppleness, and joints become more flexibile.
- It increases endorphin levels. These are natural chemicals that are effective in reducing chronic pain.

Self-massage

You can effectively massage yourself on parts you can reach. Apply a little cream or oil to your palms and use light, circular movements and long upward strokes. For immediate tension relief, you can gently knead and squeeze muscles.

Professional massage therapy

When booking a massage, make sure the therapist is properly qualified. There are many different forms of massage available, so it may help to ask the therapist what is involved. Though massage can be done while you are clothed, it is most effective on skin. During a massage, your body will be draped with a sheet, except for the area being massaged. You will only need to remove enough clothing to allow the therapist access but never more than you are comfortable with. The therapist will leave the room while you disrobe.

The most commonly taught type of massage in the United States is Swedish massage. An oil is applied to the skin, and the therapist uses a range of movements to relax the muscles; these include long smooth strokes, kneading and compressing, deep circular motions, vibration, and tapping.

The following tips may help you to get the most out of a massage:

- Tell the therapist of any painful areas so that the massage can focus on where you need it most.
- If you are uncomfortable (too cold, too warm) let the therapist know.
- If the massage is too light or too hard, ask for a change in pressure.
- Breathe deeply—and relax.

When should massage be avoided?

Massage should never be performed directly on bruises; inflamed, infected, or damaged tissues; or varicose veins.

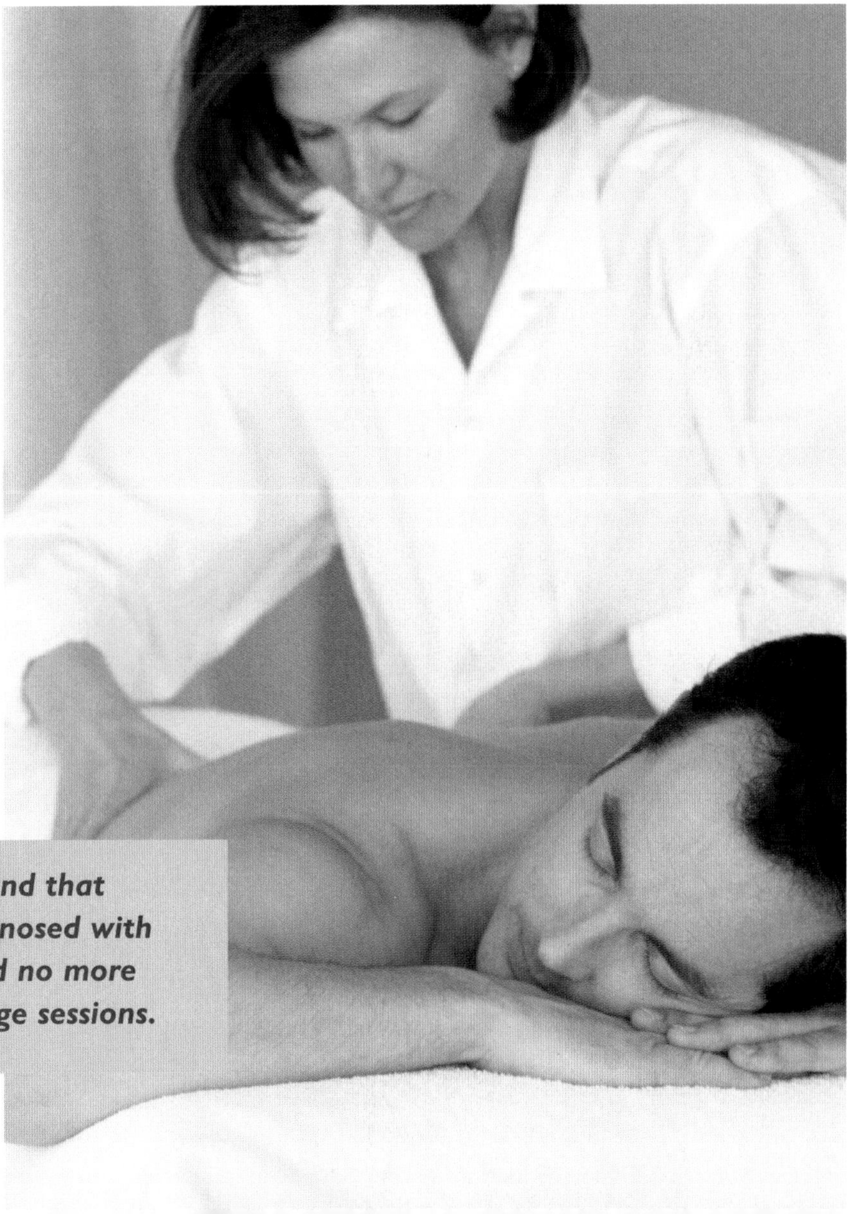

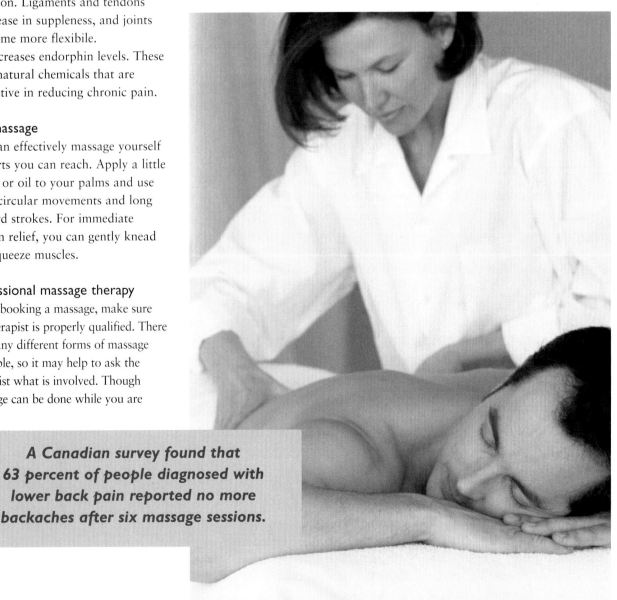

A Canadian survey found that 63 percent of people diagnosed with lower back pain reported no more backaches after six massage sessions.

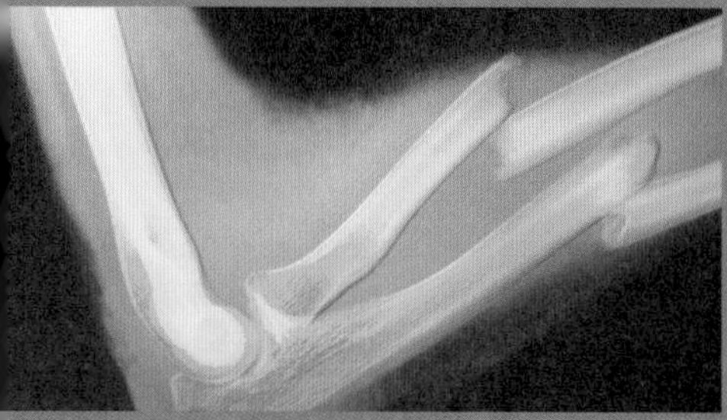  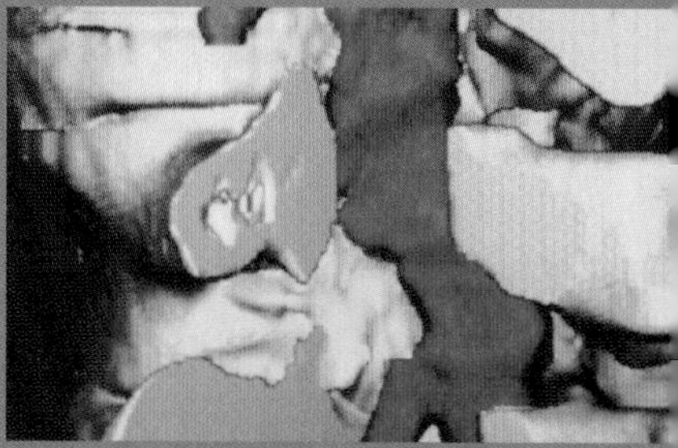

3 What happens when things go wrong

Knowing what can go wrong

Bones provide the framework that supports the body, and muscles steer and control it to power our every move. How the musculoskeletal system weathers the demands placed on it is influenced by inheritance, age, diet, and lifestyle.

AGING

Aging increases the frailty of the body's tissues for a number of reasons, a major one being that older tissues contain less water, making them less elastic, bones more brittle, and muscles weaker. Why this happens is not fully understood, but genetic factors are certainly important.

Some diseases have traditionally been considered part of the aging process—osteoarthritis in the case of stiff and sore joints, for instance. However, there is growing evidence that osteoarthritis is a distinct disease that has a different effect on the joints from age alone.

The aging process in bones and muscles is generally painless and very gradual. A normally aged joint might have reduced mobility, thinner cartilage, and weaker bone than a young joint, but symptoms of disease may only arise if other problems, such as a fracture or osteoarthritis, are added to the equation.

CONGENITAL DISORDERS

Some congenital disorders—conditions present at birth—or the musculoskeletal system are determined by genetic factors, and others are caused by outside influences, such as drugs, radiation, or other toxic

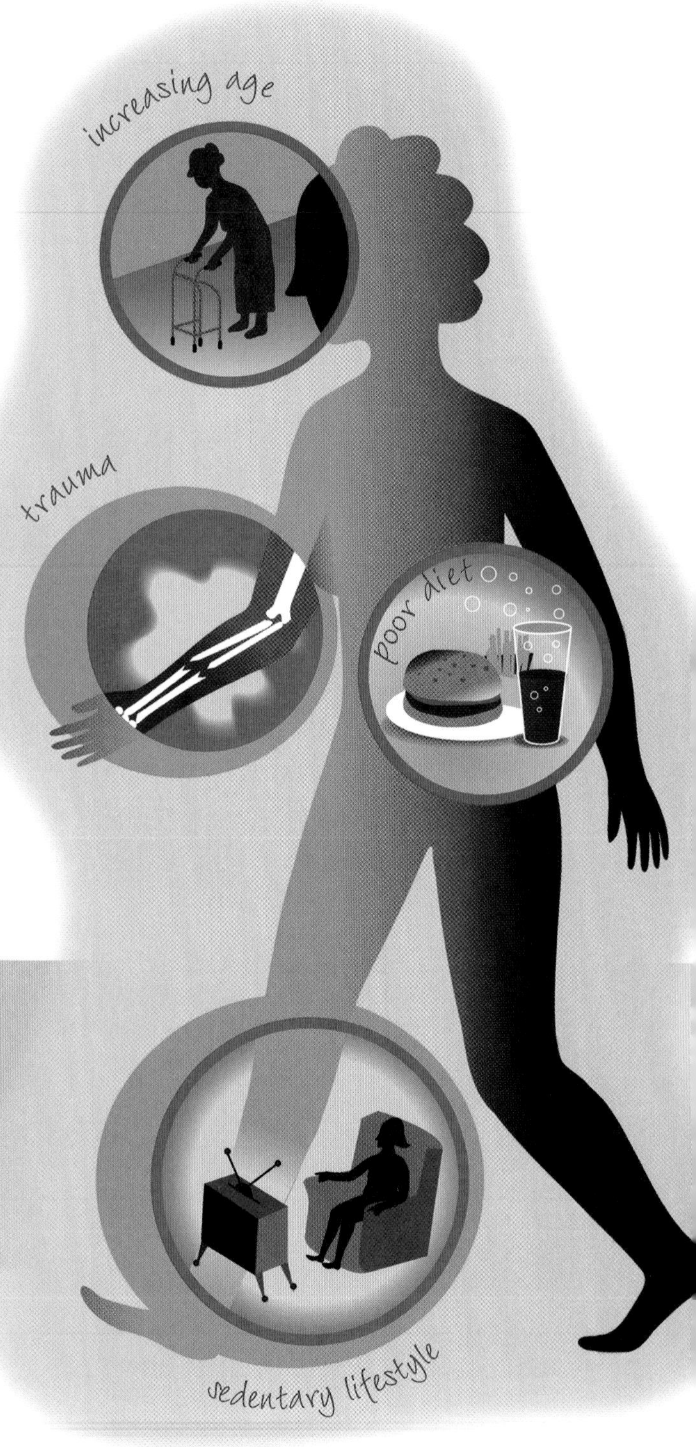

increasing age
trauma
poor diet
sedentary lifestyle

MORE COMMON

BACK PAIN
The American Academy of Orthopedic Surgeons estimates that 80 percent of Americans will experience back pain at some time during their lives. Nearly 12 million physician office visits per year are for back pain.

ARTHRITIS
About 43 million Americans suffer from arthritis and other rheumatic diseases. Rheumatic diseases are the leading cause of disability among adults age 65 and over.

The frequency of problems
Bone, muscle, and joint problems are fairly common. Some of them are highlighted here, together with recent figures indicating how common some problems are in the U.S.

substances. Types of malformation include the failure of formation of a part (arrest of development); the failure of parts to separate, as with the webbing of one or more fingers; the duplication of parts, as with more than five fingers; undergrowth; and overgrowth.

Although surgery and other methods of treatment may offer great improvement to the deformity, malformed parts are unlikely ever to be corrected completely. In addition, people with such problems are very likely to develop osteoarthritis and other complications of the affected parts. It is important to keep in mind that the functional outcome should be the priority of any treatment. People with hand deformities, for instance, often manage to perform extremely well with little or no treatment. Poorly planned treatments can lead to aggravation of the condition.

POOR DIET

A nutrient-poor diet can result in underdeveloped muscles and bones. It is particularly important to pay attention to calcium intake. The bones are both the largest user and source of calcium in the body, and if there is not enough calcium in the diet, the body may remove it from the bones to compensate. To process calcium effectively, the body needs several other substances and hormones, the most important of which is vitamin D.

Insufficient intake of vitamin D and a lack of exposure to sunshine are causes of the bone deformity known as rickets (osteomalacia, see page 145). This condition is often seen in children who are severely undernourished.

TALKING POINT

Thalidomide: still in use

Probably the best-known example of musculoskeletal birth defects triggered by an outside factor are the limb deformities of children whose mothers took the antinausea drug thalidomide. The drug was in use in Europe, Australia, and South America from 1958 until 1960 (it was never introduced in the U.S.). Since the mid-1960s, however, it has shown to be an excellent treatment for leprosy; to limit the body's immune response, thus possibly proving useful in preventing transplant rejection; to possibly limit cancer cell growth; and to reduce muscle wasting in late-stage AIDS. Birth defects are still possible in countries where thalidomide is used for these beneficial effects. Research is underway to synthesize a version that is not harmful to fetuses.

Being severely underweight is bad for the musculoskeletal system. Once the body reaches the limit of available energy in the form of carbohydrates and fats, it starts consuming the protein in its own muscle tissue and bones. This quickly leads to muscle wasting and softening of the bones, because the body robs the bones of their calcium to keep the blood levels constant. Being overweight is not particularly bad for the bones because they need loading to remain strong, but obesity does overstress joints and makes it more difficult for muscles to move.

LESS COMMON

OSTEOPOROSIS
About 10 million Americans have osteoporosis; 80 percent are women. Another 34 million have low bone mass, which puts them at risk for the disease.

PAGET'S DISEASE OF BONE
Paget's disease of bone affects 3–4 percent of Americans over age 50 and 10–20 percent of those over 60. This is approximately 3 million people. It is more common in people of European ancestry and is rarely diagnosed before age 40.

ANKYLOSING SPONDYLITIS
Ankylosing spandytitis affects 129 out of 100,000 people in the United States. It is more common in males, typically appearing during adolescence, and in Native Americans.

LIMB DEFICIENCY
Based on data from studies in the United States and Canada, it is likely that two thirds of limb deficiencies are congenital, and one third are acquired, that is, result from amputation.

The World Health Organization suggests that about 40 percent of white females will suffer an osteoporosis-related fracture in their lifetime.

UNDER- AND OVER-EXERCISING

A sedentary lifestyle is bad for both the bones and muscles. Both need to move to retain their strength and mobility. Too much time spent at a desk during the day and on the couch in the evening do not give the bones and muscles enough of a workout and—combined with a poor diet—can contribute to obesity.

On the other hand, inappropriate exercise can lead to muscle strains and sprains, to fractures, and to joint problems in later life. Exercise so intense that it causes menstruation to cease is a risk factor for osteoporosis. The best advice is to start gently if you are unused to exercise, to always warm up, and to stop an activity if it causes pain.

FRACTURES

Bone fractures are commonplace, and not just the bones are involved. There is usually damage to muscles and joints, too, and sometimes nerves, blood vessels, and internal organs, as well. Types of fracture vary considerably according to age, gender, and occupation, with the number of fractures peaking among children and teenagers and then again in old age.

- **Fractures in children** These are commonly caused by accidents associated with the roughness of childhood play (especially among boys). One characteristic of children's bones is biological plasticity: If constantly bent or forced to assume a certain position, the bones can alter shape. Therefore, simple bone fractures in children are often repaired by treatment in a plaster cast or splint, without recourse to surgery.
- **Fractures in adults** In this age group, fractures are most frequently caused by high-energy trauma, such as car crashes and sports injuries.
- **Fractures in the elderly** Bones weaken with age or osteoporosis and so break more easily. Women tend to have more fractures than men, and they are mainly of the hip, wrist, or vertebrae, often resulting from a simple fall. Recovery can take time, because bones heal less effectively in old age, so the best treatment is prevention. Simple preventative strategies include regular eyesight and balance tests; ensuring that the home is well lit; and ensuring that all surfaces underfoot are secure and stable, both inside and around the home.

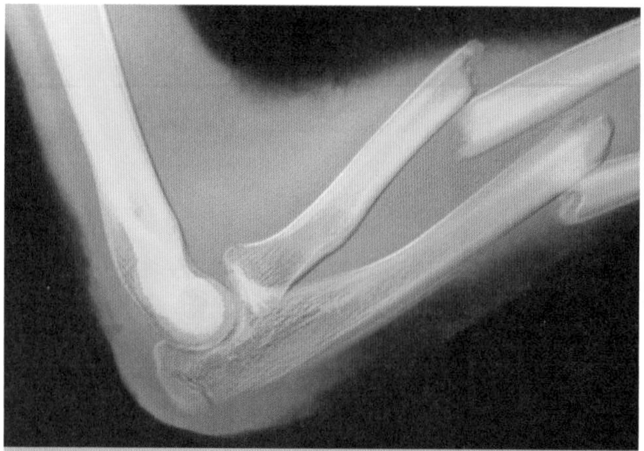

Three grades of open fracture

How badly a limb is broken is judged very much on the extent of the damage to the surrounding soft tissue. Long bone open fractures are commonly classified as follows:

Type I This is a fracture with an open wound less than half an inch long. The broken bone pierces the skin, but there is little muscle or skin crushing.

Type II This is a fracture with an open wound more than half an inch long, with a moderate degree of muscle damage and skin crushing.

Type III Wounds are more than 4 inches long, with extensive muscle and skin damage, along with injury to the nerves and blood vessels. This kind of fracture often requires plastic surgery to close the wound. These fractures are usually caused by forces of high energy such as a gunshot.

MUSCLE STRAINS

Often during sports or exercise, especially if muscles have not been warmed up properly, a twist or pull on a muscle causes a small tear. This is commonly known as a muscle strain or a pulled muscle. Strains are best treated by stopping the activity that caused the problem in the first place, then by rest, ice, compression, and elevation (RICE, see page 95).

BACK PROBLEMS

Most back pain is a result of overexertion, fatigue, or a sudden twisting or jarring movement. Poor standing and sitting postures (see page 72) are among the leading causes of back pain. If the ligaments that support and surround the discs between the vertebrae in the spine weaken, the stress of a repetitive movement such as heavy lifting or a sudden injury can cause a ligament to tear. When this happens, disc material "slips" out, pressing on nearby nerves and causing acute pain. Slipped (herniated) discs are more common in those ages 30 to 50 and in the lower lumbar spine. In those over 60, there is even more risk of slippage because the disc becomes dehydrated.

Who's who—meet the orthopedic experts

Practitioners dealing with bone or muscle problems essentially fall into one of two categories: those who manipulate the body to preserve, rehabilitate, or maximize mobility and those who are involved when surgical intervention is required.

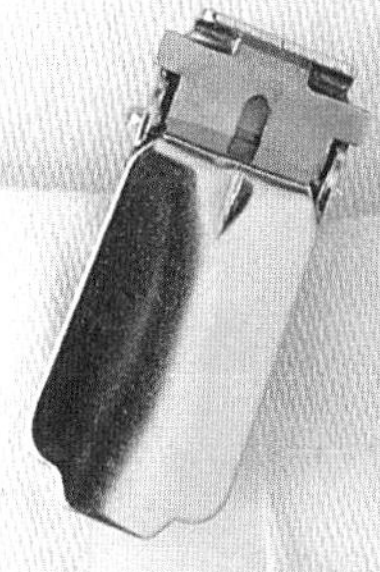

RHEUMATOLOGIST

The rheumatologist is a doctor who has become a hospital-based specialist in the treatment of degenerative conditions of muscles and joints by nonsurgical means. Drugs used in the treatment of rheumatoid arthritis can be very strong and have severe side effects. Prescription of these drugs and control of the treatment program needs an expert with specific training and a great deal of experience.

NEUROLOGIST

This is a hospital-based doctor who has specialized in the diagnosis and nonsurgical treatment of disorders involving the nervous system. This necessarily includes many conditions of the spinal cord and the musculoskeletal system. Surgery for a spinal cord condition that involves the nerves is performed by a neurosurgeon.

ORTHOPEDIC SURGEON

A hospital-based specialist concerned with surgical treatment of traumatic and degenerative disorders of the bones, joints, and muscles. Some specialize in certain areas of the body, such as knees and hips, shoulders and elbows, the spine, or the hands. Some specialize in treating children. An orthopedic surgeon works with other medical specialists and allied professionals on each case as required.

PHYSICAL THERAPIST

A physical therapist uses exercise, manipulation and massage, electrical stimulation, hydrotherapy, and a variety of other treatments to promote healing (see page 126). Physical therapy plays a key role in the rehabilitation of patients after surgery and of patients with disorders that affect their mobility. Physical therapists work in hospitals, in the community, and in private clinics. A physical therapist can be a member of the American Physical Therapy Association (APTA), but this is not required.

ORTHOTIST/PROSTHETIST

Orthotists design and fit braces, splints, and special footwear to support bones and joints that are weakened through injury or some other cause. Prosthetists design and fit artificial limbs for amputees and those born with missing limbs. Both specialists aim to maximize the comfort and physical abilities of their patients. The appliances are made by orthotic or prosthetic technicians.

PODIATRIST/CHIROPODIST

A podiatrist is a foot specialist, and podiatry is replacing chiropody as a specialty; the term is more universal. Podiatrists deal with alignment problems in the feet, knees, and back, as well as complications resulting from diabetes and sports injuries. Some podiatrists also perform surgery.

D.O. (DOCTOR OF OSTEOPATHY)

Osteopathy involves the treatment of a variety of ailments, often of bones, muscles, or joints, by means of manipulation of the relevant part of the musculoskeletal system (see page 132). Each state sets its own requirements, but D.O.s may belong to the American Osteopathic Association (AOA).

CHIROPRACTOR

Chiropractic is a system of treatment that mainly consists of manipulation of the spinal column, based on the belief that a healthy spinal column is essential for the well-being of the body as a whole. Chiropractors in the U.S. may belong to the American Chiropractic Association (ACA).

FINDING OUT WHAT IS WRONG

It may take some time for a doctor to reach a specific diagnosis of a muscle, bone, or joint problem: Several different approaches may need to be tried and their results combined to pinpoint what is wrong. Key to a successful diagnosis is a full medical history. This may be followed by one or more imaging techniques that may show what the problem is. Some conditions leave "markers" in the blood, so blood tests may be appropriate. It may also be useful to monitor the activity of a muscle or to take a biopsy of a bone or muscle before a confident diagnosis can be reached.

Medical history and examination

Regardless of the nature of a problem with the musculoskeletal system, the doctor will follow a similar series of steps to reach a diagnosis. As a typical example, the steps taken to reach a diagnosis of carpal tunnel syndrome are outlined here.

GATHERING INFORMATION

There are several steps that a doctor needs to take to reach an accurate diagnosis. The doctor will begin by taking a medical history to ascertain when the symptoms started, how they started, and if any measures or activities seem to make them better or worse.

Carpal tunnel syndrome is caused by the compression of a nerve in the wrist, which generates symptoms of tingling, pins and needles, and sometimes pain. The patient will typically complain of tingling in the affected thumb (say the right one) and the index and middle fingers, with various degrees of involvement of the ring finger.

HELP YOUR DOCTOR HELP YOU

Plan ahead!

A little forethought will maximize the effectiveness of the time you spend with a specialist doctor or therapist.

- *Arrive with time to spare before your appointment, because you may be asked to fill in forms before your consultation.*
- *Remember that you will have been allotted a specific length of time depending on the nature of your complaint. Stick to the problem at hand and don't be tempted to discuss unrelated disorders. This will distract the focus of the doctor and take away time that could have been used to deal with the main problem.*
- *Wear clothing that will allow the doctor to gain access to the affected area so that it can be examined easily and without fuss.*

Testing for carpal tunnel syndrome
Right: A hollow below the thumb indicates wasted muscles—a typical symptom of the disorder. Below: A doctor taps a patient's hand with a rubber mallet. Some doctors wear magnifying lenses to observe the response; the patient will feel the strike as a tingling.

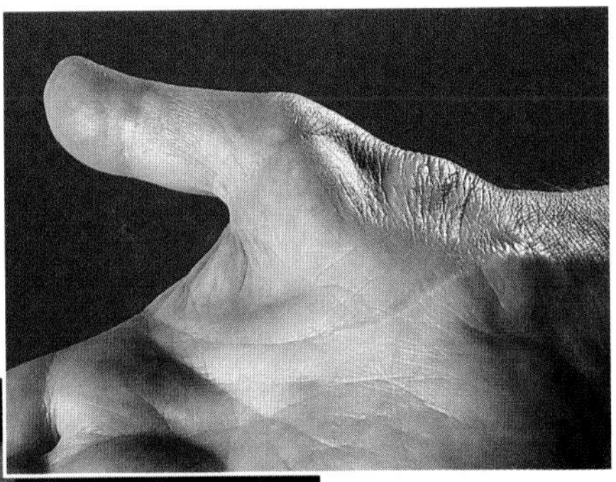

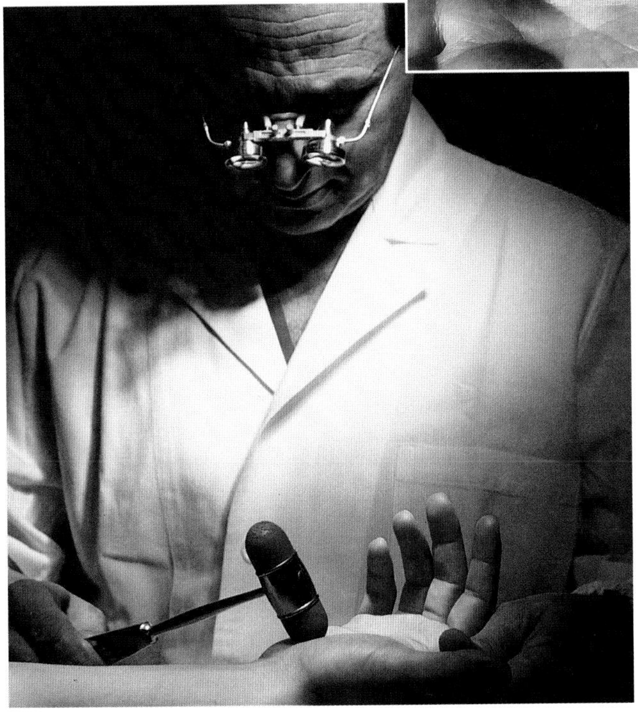

The doctor will ask about the duration of the symptoms, whether they get worse at night, if sleep is disturbed by the symptoms, and if the patient has symptoms when waking up in the morning. Further questions will aim to see if the symptoms are related to any specific physical activity, such as typing or cooking. In addition, the doctor will need to know if the pain occurs elsewhere.

Associated conditions are also important. The doctor will need to know if the patient has rheumatoid arthritis or has had a recent fracture. Both may compress the median nerve in the wrist. Finally, because surgery may be an option if other treatments (such as a brace) fail, it is important to know if the patient has any conditions that might affect the surgery, such as diabetes or high blood pressure, and what drugs (if any) the patient is taking. Even over-the-counter drugs like aspirin can affect blood clotting. The doctor will also need to know if the patient has any lifestyle habits that might affect the safety of an operation, such as smoking or excessive drinking.

LOOKING FOR TELLTALE SIGNS

After gathering this information, the doctor begins a physical examination, following a system of looking, feeling, and moving. The doctor will look to see if there are any obvious asymmetries when compared to the other hand, including muscle wasting and malformations such as a deformity resulting from a previous fracture. In the case of carpal tunnel syndrome, feeling the hand helps to assess if the muscle bulk is preserved or if there is abnormal warmth associated with inflammatory arthritis. The doctor will move the hand, or ask the patient to do so to tell if muscle strength is abnormally weak or if the range of movement of joints in the area is compromised.

As a final part of the examination, the doctor may perform other tests designed especially to investigate carpal tunnel syndrome.

REACHING A DIAGNOSIS

On the basis of taking the history, making the physical examination, and any tests carried out on the spot, the doctor will either recommend further, hospital-based tests or reach a diagnosis. With the diagnosis, the various treatment options will be outlined. If these include surgery, there are several questions the patient should ask.

- Why is this surgery being recommended?
- Are there any alternatives?
- What benefits will surgery bring?
- What are the risks involved?
- What is the success rate for the procedure?
- Is there any value in waiting to see what develops before deciding on surgery?

Do not hesitate to ask questions such as these.

- How many such procedures are performed annually at the hospital involved?
- How experienced is the doctor who will perform the surgery?
- What kind of anesthesia will be used?

Be sure to find out what will happen after an operation by asking specific questions.

- How long is the average recovery time?
- How long is required in the hospital?
- How long will rehabilitation take?
- Will physical therapy or occupational therapy be needed and for how long?

Diagnosing bone and muscle problems

In recent years, some highly sophisticated tools have been developed that enable doctors to look in detail at the bones and muscles inside the body. Such tools make it much easier for doctors to reach an accurate diagnosis and decide how best to respond.

MUSCULOSKELETAL IMAGING

In spite of the availability of more modern techniques, doctors still find that taking an X ray can be the most accurate, economical method for investigating what's going on, particularly when dealing with musculoskeletal problems such as fractures, osteoarthritis, and calcification and with benign and cancerous tumors.

X rays

X rays are invisible to the naked eye, but they blacken photographic film. If an X-ray beam is aimed at a body area, the more dense parts, like the ribs and other bones, absorb more X rays than less dense structures like the skin. This casts shadows of varying intensity onto photographic film. Dense tissue like the ribs appears white, and softer tissue like the skin appears gray.

Patients cannot feel X rays passing through them, nor does there need to be any physical contact, which helps when a patient is in pain, such as following a fracture.

Because of their high calcium content, bones appear almost opaque on X rays, making X rays an excellent way to investigate bone problems. For example, if a bone is shown to be out of alignment, it is probably fractured. Fragments of broken bones will show up as spikes.

An X ray of osteoarthritis of the spine

The vertebrae of the lower back of a patient with osteoarthritis. This disease is common among older people: 80 percent of those over 65 show evidence of its presence on X rays, although only a quarter of these will have symptoms, including pain.

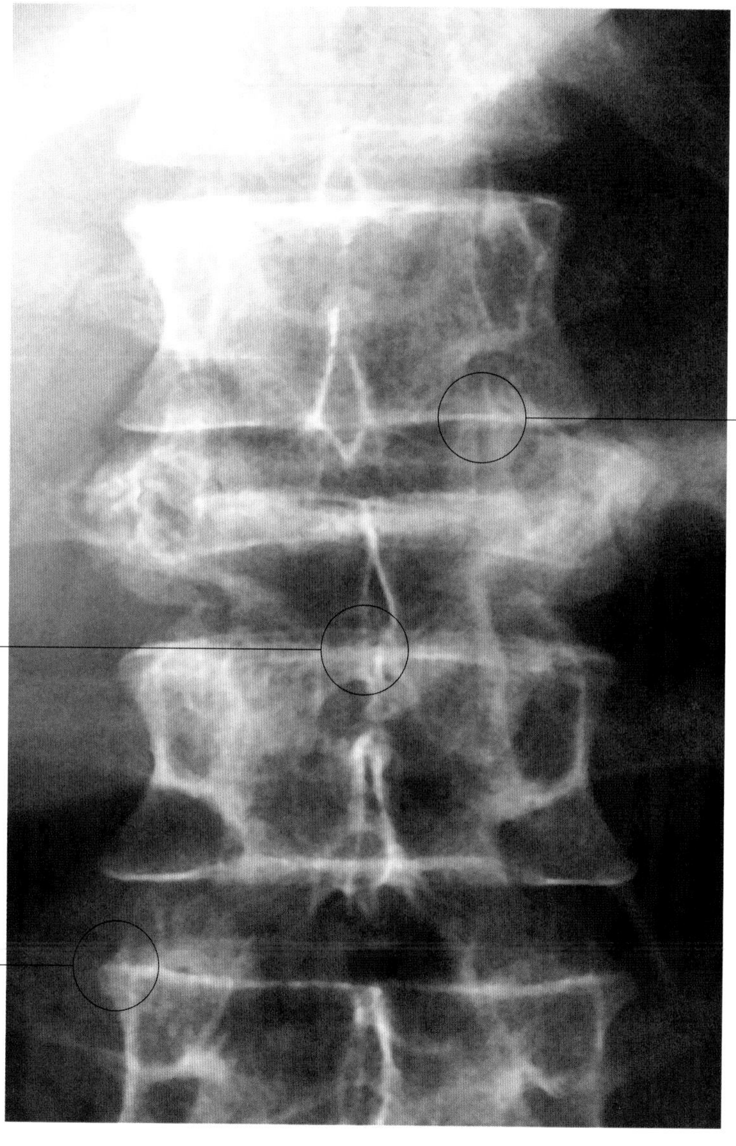

BONE UNDERNEATH CARTILAGE SHOWS WHITER THAN OTHER BONE
The bone underlying the cartilage becomes distinctively whiter because of repeated microscopic fractures of the bone, which makes it look similar to a hardened sponge.

OSTEOPHYTES APPEARING AROUND THE JOINTS
As the condition becomes more severe, bony spikes (osteophytes) appear around a joint.

REDUCTION IN SPACE BETWEEN JOINTS
The joints, the shape of the bones, and the space between the bones determine the extent of osteoarthritis. In this condition, the cartilage is usually affected before bone, so a reduction in joint space as the cartilage shrinks is often the first radiological sign of the condition.

CYSTS IN THE BONE (not visible)
Cysts in the bone, resembling round perforations, indicate that the bone is no longer able to sustain the load it has to carry.

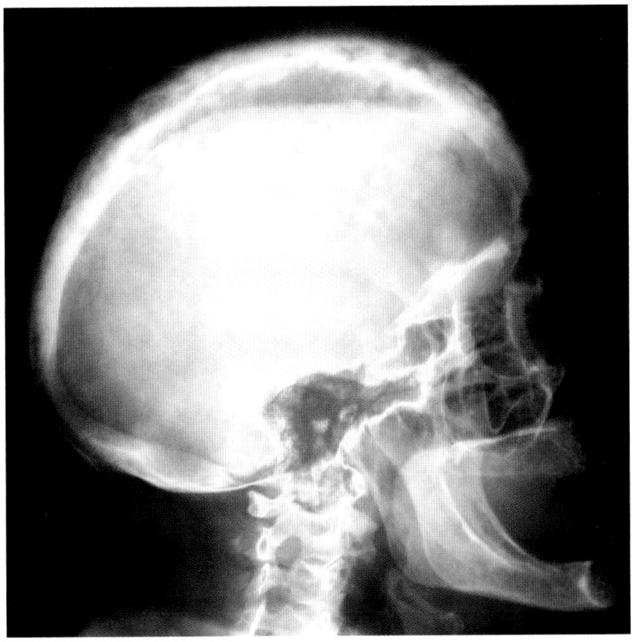

X-ray evidence of Paget's disease of bone

In Paget's disease, excessive bone renewal leads to bone becoming thicker yet weaker and therefore more likely to fracture (see page 147). In this X ray, the disease is indicated by the mottled appearance of the bone's surface, showing increased porosity, and the thickening of some skull bones, giving a grayish brown coloring.

X rays seldom show important information about soft tissues, but certain features, such as gas or air in the tissues, may show up, and these can indicate a severe infection or an open wound.

Ultrasound scanning

Sometimes, when trying to diagnose a musculoskeletal problem, a doctor will ask for an ultrasound scan. Undergoing an ultrasound scan causes no discomfort and is considered totally safe. A transducer is placed on the skin, sometimes with gel to improve the contact, and transmits inaudible high-frequency sound waves into the body. Structures inside the body reflect these sound waves back as echoes. The transducer picks up these echoes, which are converted to numerical data and displayed directly on a screen or analyzed by a computer to produce an image. Ultrasound is useful in diagnosing musculoskeletal disorders because it can detect tears in muscles and collections of pus or blood within the tissues.

Computed tomography (CT) scans

There are times when it might be difficult for a conventional X ray to pick up a crucial detail, such as when dealing with the small bones of the hand, where there are eight bones closely held together in the wrist. Fractures here often go unnoticed even by the most experienced practitioner. Or perhaps the doctor needs to examine a feature that is hidden by several layers of tissue. The solution may be to have a CT scan.

In CT scanning, several X-ray beams are passed through the body and then analyzed by a computer to reconstruct a cross section of "slices" through the body. In cases such as a complex hip or shoulder fracture, CT scans can provide the surgeon with a three-dimensional picture showing where fragments of fractures are located.

Magnetic resonance imaging (MRI)

For an MRI scan, the patient lies in a hollow cylindrical magnet and is exposed to a strong magnetic field. The force of the magnetism causes protons in the water in the body temporarily to line up parallel to each other (they normally point in random directions). A strong pulse of radio waves knocks the protons out of alignment, and as they realign themselves, they send out a radio signal that is captured by the sensors and used to construct an image.

At first glance, MRI scans look very much like CT scans, but they give a far more detailed picture of soft tissue because of the differences in water content within the tissues. MRI scans can give an image of almost any

Are X rays safe?

Frequent exposure to X rays can damage the skin and cause cancer, but exposure is now so low that the risk is very small. Initially, a high amount of radiation was needed to obtain an X-ray image on photographic film. As the decades passed, however, techniques were developed that greatly decreased the amount of radiation needed to capture an image. These have included the use of more sensitive films; the creation of fluorescent and phosphorescent tubes that emit light when hit by X rays and amplify the effect; and, most recently, the development of electronic detectors that measure the intensity of the X ray to produce an image on computer.

ASK THE EXPERT

human tissue and they are particularly useful for examining the brain and spinal cord. Using different techniques, MRI scans can detect fractures and soft-tissue problems such as tears in the cartilage, which previously could only be seen through surgery.

Diagnosis through MRI is not 100 percent accurate, and sometimes it may be difficult to visualize small cartilage injuries. Also, because of the strong magnetic fields, this diagnostic tool is not suitable for patients with plates, screws, pacemakers, and other metallic or electric devices in their bodies.

Radioisotope bone scanning

This method of investigating the condition of the bone consists of injecting substances that have a low radioactivity, such as technetium, into the blood and then detecting them using equipment similar to Geiger counters. If there is an increase in blood flow, such as when fractures are healing or when there is an infection, the uptake of the substance and ensuing radioactivity will be higher. Conversely, if the bone is dead (necrotic) or has healed, the uptake of the substance will be reduced or nonexistent. This method is often the most sensitive way to detect alteration in the bone, including the presence of cancer that may have spread from another site. The disadvantage is that it often fails to discriminate between possible causes; therefore, other methods of investigation, such as X rays, MRIs, or even biopsies may have to be used as well.

MEASURING BONE MASS

As a tool for diagnosis, measurements of bone mass—bone mineral density—give information on the presence or absence of disease and can also be used to predict how fragile the bone is likely to become in the future. The most frequently used method is dual X-ray absorptiometry (DXA). Other methods, such as ultrasound of the heel bone and computerized analysis of wrist and hand X rays, are also reliable measurements of bone mass.

BONE MARKERS

Bone markers are proteins or other substances associated with bone metabolism that can be measured in either blood or urine samples in order to give information concerning the presence or absence of disease. Bone, like any other living tissue, is constantly being renewed: New bone is being formed, and existing older bone is being resorbed. The concentration of calcium and hydroxyproline in the urine, for instance, can give an estimate of the amount of bone absorption taking place.

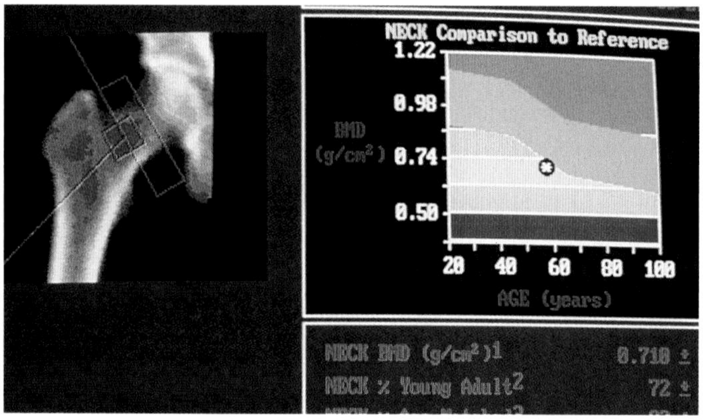
NECK Comparison to Reference
1.22
0.98
BMD (g/cm²)
0.74
0.50
20 40 60 80 100
AGE (years)
NECK BMD (g/cm²)1 0.710
NECK % Young Adult2 72

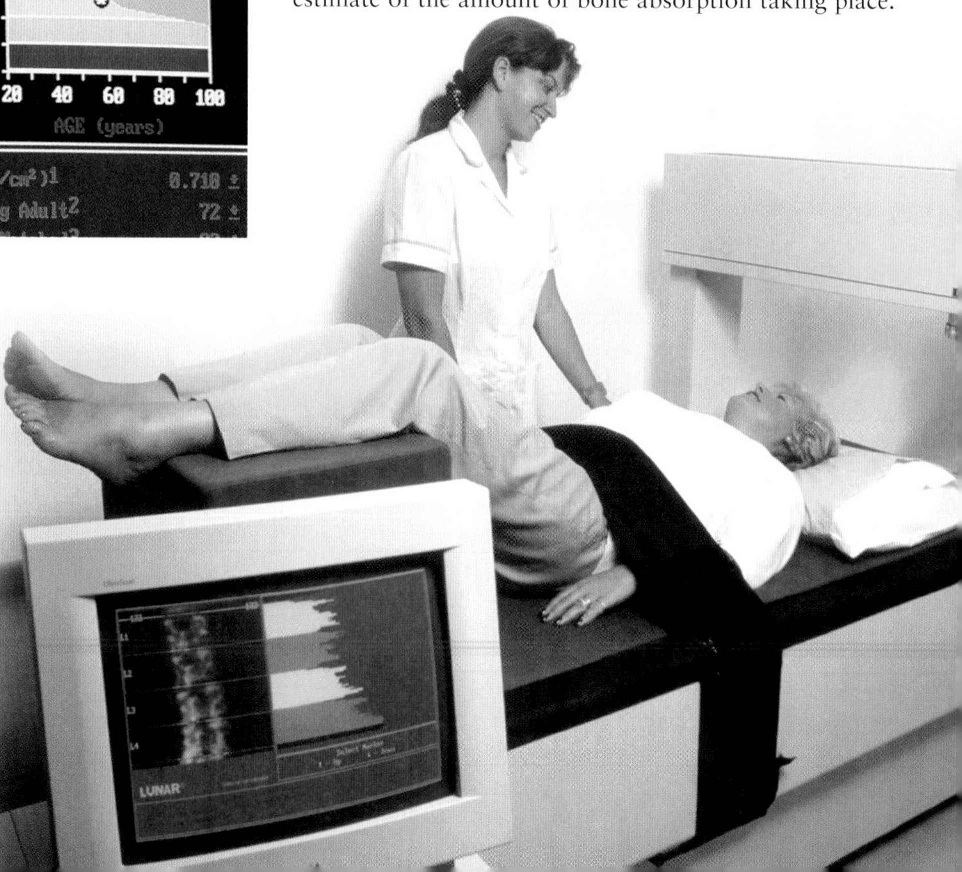
LUNAR

Measuring bone mass

Right: This patient is being tested for bone mineral density using dual X-ray absorptiometry (DXA)—X rays and an electronic sensor are used to analyze the bone mass (density) at every point in the bone under investigation. Above: A monitor image shows the direction of a scan being taken through the neck of the femur, and the graph next to it indicates the average age at which bone density generally falls to the level observed in this patient.

Osteoporosis in color

One of the most effective tests for indications of osteoporosis is a measure bone mass, which is then color-coded on a false-color scan.

a This scan of bone density shows a spine free of osteoporosis. The densest bone is seen as red, then less dense bone as yellow and blue, with least dense bone colored green.

b This scan shows the spine of a woman with osteoporosis. It gives quite a different picture, with bone density running from blue to green and no red or yellow showing at all.

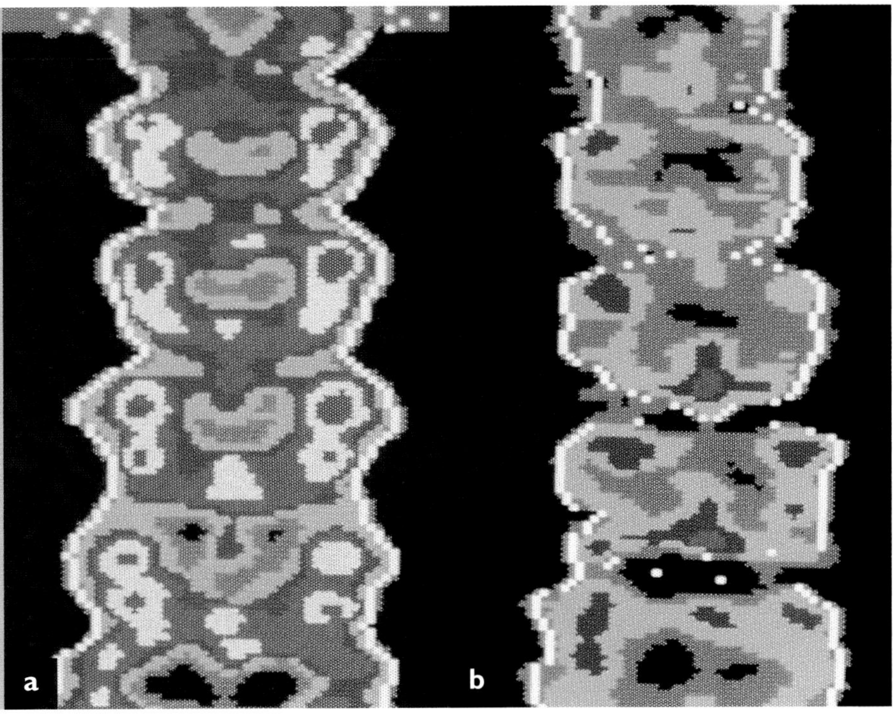
a
b

BLOOD TESTS

Blood tests are usually not specific enough to provide a diagnosis but are a good screening tool that can yield much useful information.

- **Erythrocyte sedimentation rate (ESR)** This is the rate at which red blood cells settle out of suspension in plasma. An hour after blood is taken, the amount of clear plasma in the test tube is measured. ESR is faster in inflammatory conditions such as rheumatoid arthritis and systemic lupus erythematosus (SLE).
- **C-reactive protein (CRP)** CRP is normally present in small amounts but is released in greater quantities in response to injury, infection, or inflammatory conditions.
- **Hemoglobin** The blood of patients with rheumatoid arthritis or SLE may show that they are anemic.
- **White blood cell count** The white blood cell count is elevated in infections such as septic arthritis.
- **Uric acid** This will be measured if gout is suspected.
- **Enzymes** Creatine kinase, a muscle enzyme, may be raised in inflammatory conditions; alkaline phosphatase is raised in osteomalacia and Paget's disease of bone.

ELECTROMYOGRAPHY

Electromyography records a muscle's electrical activity both at rest and while it is contracted. The technique can be used to determine whether a muscle problem is caused by weakness or disease in the muscle itself or by problems in the nerves supplying it. One or more needle electrodes are inserted through the skin into the muscle. The electrodes measure muscle activity, with the result displayed on an oscilloscope as an electrical waveform; the "sound" of any muscle activity may also be measured (healthy muscle at rest should be silent). The procedure provides information about the ability of the muscle to respond to stimulation. Conditions that cause abnormal results include inflammation of the muscle tissue, muscular dystrophy, spinal cord injury, multiple sclerosis, and diabetic neuropathy.

BIOPSIES

A muscle biopsy may be taken to ascertain whether there is any inflammation or degeneration and whether a problem is caused by the muscle itself or affects its nerve supply. A sample is taken in a minor operation under local anesthetic. Alterations in the size, shape, and distribution of the muscle fibers help the doctor reach a diagnosis.

A bone biopsy may be performed under local or general anesthetic. The latter means that the bone sample can be analyzed immediately and, if necessary, a tumor can be removed. A biopsy allows the doctor to differentiate benign from malignant tumors and may also indicate the presence of rickets, osteomalacia, and infection.

CURRENT TREATMENTS

There are several approaches to treating bone and muscle problems. Many conditions respond well to medication and rest. When rest is not enough, some form of support might be necessary to immobilize a bone while it heals. Surgery on the joints can offer dramatic results: Joint replacement—introduced in the 1960s—is now a common and successful treatment for many painful conditions. Finally, a patient with a musculoskeletal problem may be offered a "hands-on" treatment such as physical therapy or osteopathy.

Drugs for bone and muscle problems

Drug therapy for the musculoskeletal system is largely aimed at reducing inflammation, relieving pain, relaxing muscle spasms, and preventing and treating the potentially crippling bone disease osteoporosis.

TREATING OSTEOPOROSIS

There are currently four major drug treatments for osteoporosis: calcium and vitamin D supplements; hormone replacement therapy (HRT); selective estrogen receptor modulators; and bisphosphonates. These treatments are used at different stages of the disease and have different indications. None should be taken long-term without medical supervision: All have side effects.

Calcium and vitamin D supplements

Calcium supplements are only required when dietary intake is low. In certain conditions, such as pregnancy and lactation, infancy, and old age, requirements increase. With osteoporosis, an intake of double the recommended daily allowance may reduce the rate of bone loss. Calcium salts are available in several oral preparations, including chewable or effervescent tablets. However, calcium should be avoided by people with kidney problems and those with high levels of blood calcium. If any side effects—such as gastrointestinal disturbances, slowed heart rate, or arrhythmias—are experienced, the supplements should be discontinued and the situation discussed with a doctor.

Vitamin D deficiency can usually be prevented or treated by taking a low dosage oral supplement, such as 10 to 20 micrograms of ergocalciferol, although people with intestinal malabsorption or chronic liver disorders may require vitamin D in higher doses. Newer vitamin D derivatives such as alfacalcidol and calcitriol can be given to patients with renal problems. Symptoms of overdose include lassitude, nausea, vomiting, and diarrhea. If any of these symptoms occur, the drug should be immediately stopped and a doctor contacted.

Hormone replacement therapy (HRT)

HRT is the general term for hormonal supplementation during and after menopause, when levels of progesterone and especially estrogen fall. HRT affects several body systems.

Estrogen decreases bone loss. Extrogen receptors on bone-forming cells block the effect of PTH on those cells and prevents them from signaling bone-resorbing cells. It also has an indirect effect in that it increases calcium absorption and reduces calcium loss from the kidneys. Because estrogen deficiency is associated with subsequent bone loss, HRT should be offered to all women with premature menopause. In women with an intact uterus, estrogen treatment alone results in an increased risk of endometrial cancer, so a combination of estrogen and progesterone is used.

HRT's beneficial effects in combating osteoporosis is an important factor in its favor. Long-term estrogen replacement, however, generates several potential problems, notably increased risk of breast cancer, intravenous thromboembolism, stroke, endometrial cancer, and (some studies indicate) heart disease.

Selective estrogen receptor modulators

Drugs in this class, such as raloxifene (Evista), have a positive effect on bone metabolism, inhibiting resorption of bone, preventing bone loss, and reducing vertebral fractures in postmenopausal women. Unlike HRT, these drugs have no effect on menopausal symptoms such as hot flashes.

Bisphosphonates

When taken by women after menopause, bisphosphonates act to counter osteoporosis by reducing the rate of bone loss. However, when taken in large doses or over prolonged periods of time, bisphosphonates, in particular disodium etidronate, may actually reduce or impair bone mineralization. This means that these drugs can be effective in the treatment of Paget's disease (see page 147). Doses and duration of treatment must be carefully controlled. Bisphosphonates may also help reduce high levels of blood calcium in certain cancers. Risedronate sodium and alendronic acid are prescribed to treat osteoporosis.

Some 50 studies, involving a total of 12,000 women, have demonstrated that raloxifene increased bone density by 2 percent in two years.

ANTIINFLAMMATORIES

Antiinflammatory drugs block the production of prostaglandins, chemicals that contribute to pain and swelling in inflamed muscles and bones. In some conditions, notably rheumatoid arthritis, the inflammatory reaction starts without a clear cause. Gout is inflammation caused by the deposit of uric acid crystals, but why this happens is still unclear.

Antiinflammatories are relatively unspecific, acting on the symptoms and not necessarily the cause of the inflammation. Reductions in pain, redness, and swelling are the main benefits, regardless of the causes. There are two main classes of antiinflammatories: corticosteroids (commonly called steroids) and nonsteroids (NSAIDs).

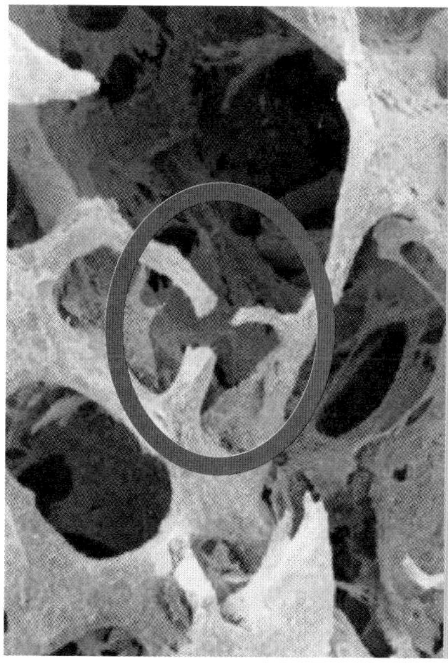

How bone breaks down

As osteoporotic bone thins and the narrow bars of bony tissue eventually break (area circled, left), making the bone fragile and liable to fracture.

a Cancellous bone with normal density and structure.

b In osteoporosis, the bone initially starts to thin.

c Eventually the bone loses its connectivity and gaps appear.

d After drug treatment, bone density increases, but the bone may still be weak because of structural damage.

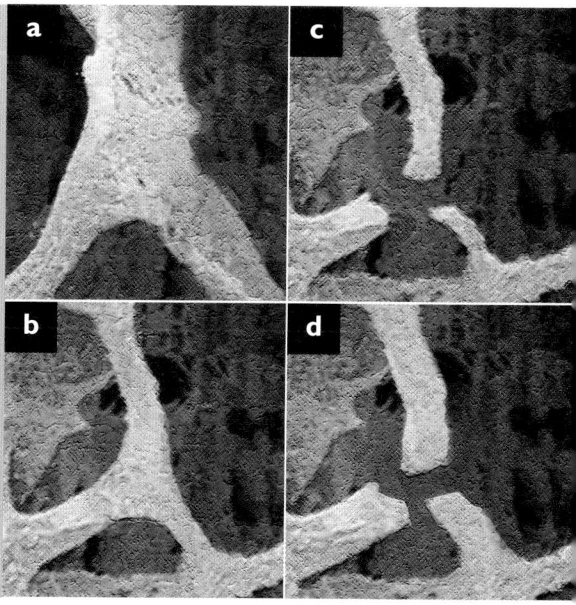
a
c
b
d

Steroids

Steroids have a very strong antiinflammatory effect. Unfortunately, their side effects are considerable, so their use is generally restricted to conditions such as severe rheumatoid arthritis, where deformity and effects of the disease outweigh problems that may come with the steroids. Side effects can include hypertension, water retention, potassium loss, diabetes, osteoporosis, muscular wasting, and even mental disturbance. Prednisone is the most frequently used oral steroid.

With tendinitis, steroids can be given in relatively high local concentrations with few or no systemic side effects by injecting the steroid into tissue close to the tendon or in the bursa (the membrane around the tendon). This type of preparation can also be injected into joints and can give substantial relief for prolonged periods of time from conditions such as osteoarthritis. There are some toxic effects on cartilage, so this treatment is used judiciously.

Nonsteroidal antiinflammatories (NSAIDs)

NSAIDs are available as tablets, suppositories, creams, and injections, although the injectable form is rarely used outside hospitals. When used regularly in appropriate doses, NSAIDs have a lasting antiinflammatory and analgesic (painkilling) effect. The lasting analgesic effects and enhanced antiinflammatory properties of NSAIDs such as aspirin, ibuprofen, and piroxicam make them the preferred choice in conditions such as osteoarthritis and rheumatoid arthritis. There is a long list of NSAIDs of relatively similar strength and properties, but the tolerance of individual patients varies, and therefore it may be necessary to try several different NSAIDs if they are needed on a long-term basis.

An NSAID may require up to a week to achieve a full analgesic effect and up to three weeks to achieve its full antiinflammatory effect. They should be used with caution in the elderly, in patients who have allergic disorders, and in particular by patients who have had previous episodes of hypersensitivity to aspirin. NSAIDs are not given to patients with active peptic ulcers and are stopped if gastrointestinal symptoms, particularly bleeding, develop. Side effects are rare, however, and NSAIDs are considered safe enough to be sold without a prescription, but courses longer than two weeks or doses higher than those indicated on the packaging should be discussed with a doctor.

Topical NSAIDs are available but not common, and their effect, particularly long term, is doubtful because it is difficult to make dosages high enough. However, they can give some relief to superficial sprains and bruising.

Flower power
Preparations from the seeds and roots of the autumn crocus, or meadow saffron, are a traditional treatment for the symptoms of gout and rheumatism.

MUSCLE RELAXANTS

Muscle relaxants are used to treat muscle spasms that have not been relieved by an NSAID such as diclofenac, or a painkiller such as acetaminophen. Taken regularly, muscle relaxants can reduce muscle stiffness and improve mobility.

Muscle relaxants act in different ways. Dantrolene acts directly on the muscle by reducing the response of muscle cells to nerve signals needed to cause the muscles to contract. Tizanidine acts on receptors in the blood vessels, stimulating the nerves that cause muscle relaxation. Botulinum toxin, when injected locally, acts directly on the nerves to block the passage of nerve signals that trigger muscle contraction. Centrally acting drugs, such as baclofen and diazepam, reduce transmission of the nerve signals from the brain and spinal cord that usually cause muscle contraction. Reducing nervous stimulation allows the muscles to relax.

DRUGS FOR SPECIFIC PROBLEMS

Gout can benefit in the acute phase from colchicine and in the chronic phase from allopurinol and probenecid, which help reduce the formation of uric acid or increase its excretion in the urine. Gold salts, penicillamine, and immunosuppressants can help to control rheumatoid arthritis, but they do have side effects.

Living with arthritis

If you have arthritis, life can be a challenge, but there's a lot you can do to help yourself with diet and exercise. In addition, modifications around the home and garden and devices designed to make life easier can lessen the condition's impact.

Your doctor can prescribe medication to alleviate arthritis pain, but self-help measures are equally important.

DAILY LIVING

Conserving energy can minimize pain and fatigue and make it easier to do the things you want or need to do.

- Plan ahead: Organize "must do" tasks so that periods of activity can be interspersed with rest periods.
- Do the minimum—take the car to the carwash, iron only the essentials—and use every available labor-saving device such as a frost-free freezer and a dishwasher.
- Organize storage so that you are not reaching or bending for frequently used equipment.

DIET

A healthy diet based on whole-grain starchy foods, fruit, and vegetables, and low in fat and sugar, will keep you around an ideal weight and thus take strain off your joints. Even a small amount of excess weight puts unnecessary pressure on weight-bearing joints. In addition, essential fatty acids help some people with arthritis. Essential fatty acids are found in fish oils (in mackerel, sardine, and pilchards, for example), in plant seed oils, and in evening primrose oil. Aim to eat three or four servings of oily fish each week.

EXERCISE

Exercise contributes to weight loss, increases strength and suppleness, and keeps joints mobile. The key is to do what you can: Walking is usually beneficial, and swimming and cycling may not stress weight-bearing joints.

A helping hand
Use any available aid, such as a walking stick, to help get out and about. A regular walk in the fresh air brings many benefits and lifts the mood.

HELPFUL GADGETS

These lists are by no means exhaustive, but they give an idea of the devices available that may help with day-to-day activities.

- **Improving grip and leverage** "Easygrippers" to hold toothbrush, knife and fork, razor; large-handled can opener, vegetable peeler; angled knives with easy-grip handles; handled electric plugs (also consider relocating frequently used sockets higher up the walls); lever taps; lever appliance knobs; lever doorknobs (or fit-extended handles to round knobs); Yale-key extension; cord-pull light switches.
- **In the kitchen** Lightweight saucepans with large heat-resistant handles; jug kettle tipper.
- **Extending your reach** Long-handled dustpan and brush; long-handled gardening tools; "reacher" for drawing curtains and picking items off the floor.

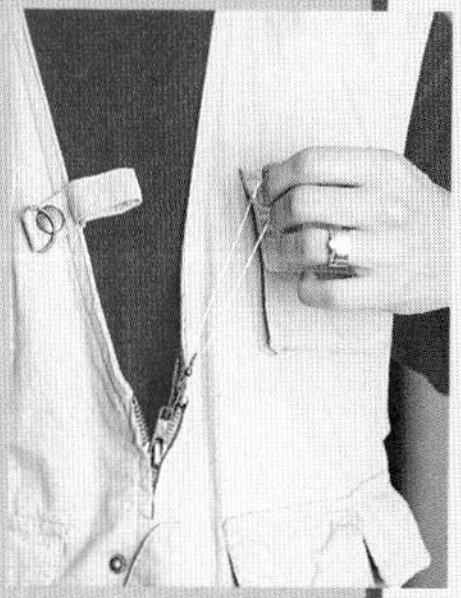

- **Help with dressing** Elastic shoelaces, long-handled shoe horn; velcro-closing underwear; dressing stick, buttonhook, zip pull.
- **Getting around** Walking stick with padded handle (for comfort and to absorb shock) and wrist strap; easy-reach seatbelt, padded steering wheel cover, panoramic rearview mirror, blind-spot mirrors, padded headrest.

Laws vary by state, but many states require physicians or others to report disabilities that can impair driving ability to that state's Division of Motor Vehicles (DNV). The DMV may then reevaluate the driver.

Support for injured bones and joints

The types of devices used to treat and stabilize fractures and other injuries vary greatly. There are those that give external support—notably casts and splints—and in addition, there are those that are implanted surgically, such as plates and screws used to fix fractures.

After injury or surgery, bones, muscles, ligaments, and tendons need long-term rest and support so that the healing process can take place unimpeded. This is best achieved by immobilizing the parts concerned.

WHEN IS IMMOBILIZATION APPROPRIATE?

There are a great many conditions for which support and immobilization are prescribed. These include

- fractures (see page 139),
- sprained or torn ligaments (page 143),
- strained muscles and ruptured tendons (muscle strains, page 144; ruptured tendons, page 151; sports injuries, page 153),
- whiplash (page 155),
- dislocated joints (page 138),
- tendinitis (page 154),
- postoperative support for limbs and postoperative protection for tendon and ligament repairs,
- for prevention of deformities following surgery, and
- for correction of deformities such as club foot (page 116).

HELP YOUR DOCTOR HELP YOU

The dos and don'ts of plaster and synthetic casts

- ***Do** counteract swelling by elevating the limb above heart level as much as you can.*
- ***Do** gently exercise fingers and toes if possible to encourage the circulation and discourage swelling.*
- ***Do** consult your doctor if the skin around the cast becomes sore.*
- ***Don't** get a cast wet if the cast itself or its padded lining is not waterproof.*
- ***Don't** be tempted to poke anything into the cast to scratch an itch; try cold air from a hair dryer instead.*

EXTERNAL SUPPORTS

The materials used to make a cast, splint, or other support should be hard enough to support the limb but soft enough not to cause further complications. This is achieved by a hard outer shell made of a synthetic material or plaster and a soft, padded lining.

Casts

A cast is molded to the individual limb. It is most frequently used to immobilize a broken bone. Fiberglass and plaster casts are both commonly used in the United States. A third type, thermoplastic resin casts, are not. Each has advantages and disadvantages.

- **Plaster of Paris casts** Casts made of gauze bandages embedded with plaster of Paris (powdered calcium sulphate, obtained by heating gypsum) were first devised in the mid 19th century. Plaster of Paris casts can be more easily, quickly, and cheaply molded to the patient's anatomy than the newer synthetic casts and so are still popular with many surgeons, especially when the casts must be molded particularly closely to the patient's limb or when a cast is needed for a few days only, as after some operations, for example. On the downside, plaster of Paris casts are heavier than the synthetic alternatives, and the patient must be very careful to keep them dry at all times.
- **Fiberglass casts** In the 1970s, the development of fiberglass casting tape made possible the production of a cast that was harder wearing and lighter than its plaster of Paris equivalent. Although this cast is more expensive, these qualities make a fiberglass cast a popular choice for long-term use. A fiberglass cast is water resistant, but its padded lining is not, so a fiberglass cast must still be kept dry. Only padding made of Gortex is waterproof, and Gortex is not a suitable choice for every injury.
- **Thermoplastic resin casts** These are not as commonly used as fiberglass casts. A plate of plastic is softened by heating in hot water and then is molded to the shape of the injured limb before the plastic cools and hardens. The molded cast is held in place with Velcro straps.

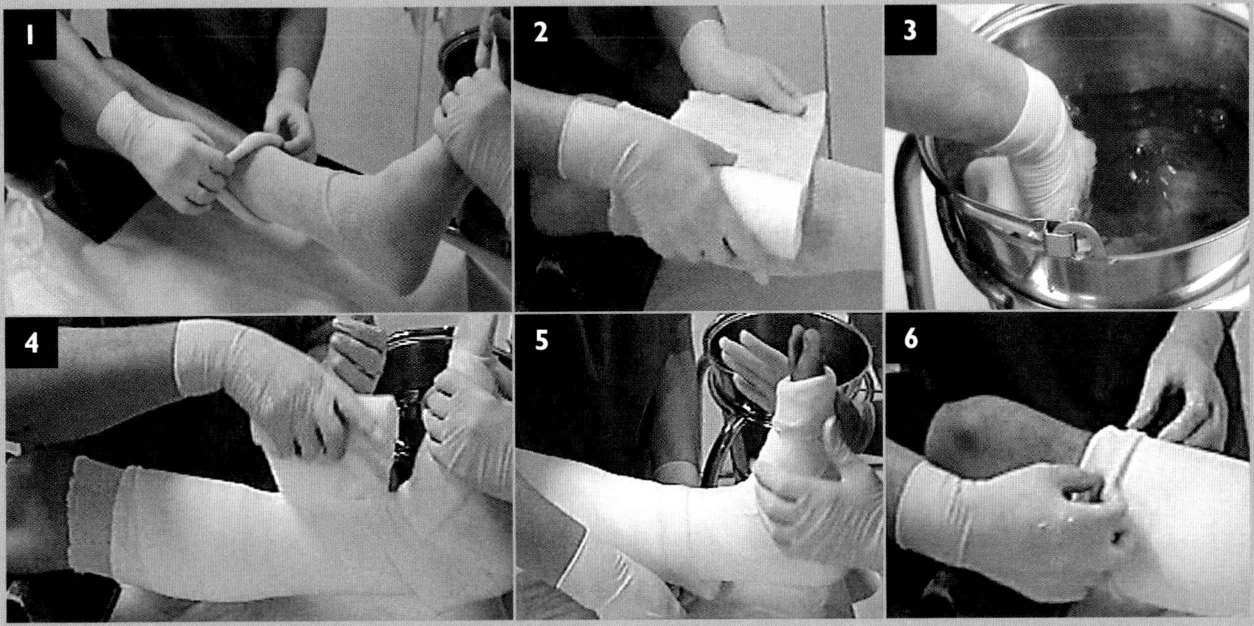
1
2
3
4
5
6

Fitting a fiberglass cast

The procedure for fitting a fiberglass cast is similar to that for fitting a cast made of plaster of Paris. Both involve wrapping the limb in wet bandages that then harden to form the cast.

1 The patient is being fitted with a cast that will cover the calf and ankle—typical for a tibia or fibula fracture. With the knee elevated and supported, the nurse or technician covers the skin with a protective layer of stockinette.

2 The stockinette is wrapped in a thick layer of cotton wool padding.

3 A bandage impregnated with fiberglass is immersed in water. The water makes the bandage soft and pliable, but the bandage will harden in just a few minutes.

4 The wet bandage is unrolled and wrapped around the patient's leg, ankle, and foot. Bandaging in this way enables the cast to be molded to the limb.

5 To provide maximum stability for the healing bone, the cast holds the ankle joint rigid. Only the toes protrude.

6 The exposed edges of stockinette are turned over at the top and bottom of the cast to protect both the patient's skin and the edge of the cast.

7 Twenty minutes later, the new cast has hardened and the patient is ready to leave. Often, simply standing on a leg encased in a cast will apply force in the right direction and so actively promote the healing of a broken bone.

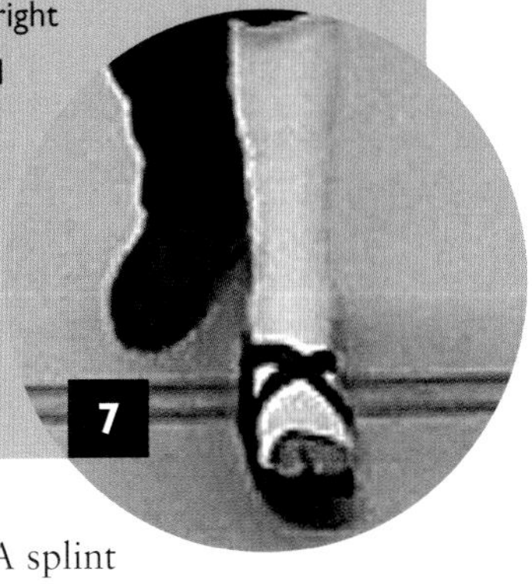
7

Splints

A splint consists of a rigid support plus a means of attaching the support to the damaged limb. It differs from a cast in being "off the shelf" (usually), offering less protection to the injured body part, and in being easier to assemble and remove.

- A splint as a first aid measure Splinting is often used to immobilize an injured arm or leg immediately after an injury to reduce the chance of the injury worsening. A splint can be improvised from a piece of wood or even a rolled-up magazine tied to the limb with bandages.
- A splint as a treatment option A splint can be used to immobilize a dislocated joint while it heals and also for some closed fractures. Splints are made of various materials, including acrylic, polythene, aluminium, and plaster of Paris. They often have Velcro straps for easier adjustment. Sometimes, especially with babies, adhesive strapping is used to form a malleable splint. Some hand splints incorporate an elastic band or a spring that allows finger movements and so prevents cut tendons in the palm of the hand from permanently stiffening as they heal. There are times

when a removable splint is the preferred choice, such as when treating tendinitis. The splint can be taken off from time to time to allow for a certain amount of exercise, which helps reduce muscle wasting and joint stiffness.

Other forms of external support

- **Collars** A collar provides extra support for neck injuries, and may also help to relieve pain by restricting movement of the head. After an accident, a collar is often used to immobilize the neck on a "just-in-case" basis, when there is a possible head or neck injury.
- **Braces** These perform much the same function as a splint. Sometimes they provide permanent support for a disabled limb or other body part rather than short-term support as part of a treatment program.
- **Slings** These are used to support an injured arm, often together with a cast or a splint. A sling usually takes the form of a triangular bandage placed under the arm and tied around the neck.

CONTINUOUS PASSIVE MOTION

The observation that movement is beneficial for a limb that is otherwise immobilized has led to the invention of the continuous passive motion (CPM) machine. This device consists of a splint attached to an electric motor. The splinted limb is kept in continuous, gentle motion for the duration of the treatment session. The range and speed of movement can be adjusted to suit individual needs. Continuous passive motion is particularly effective for fractures inside joints, because the movement promotes healing by encouraging new fibrocartilage to fill the cracks in the joint.

TRACTION

Traction is used both to realign fractured bones into their correct position and as a form of support and immobilization. There are two types of traction: continuous and intermittent. The former involves surgery: A pin is put through a bone, pulleys and weights are attached to the pin, and then continuous pressure is applied to move the bone. This may be done, for example, to treat a fracture of the femur by realigning the two parts of the bone then holding them together against the tension of the thigh muscles so that they heal correctly. Intermittent traction, which does not need surgery, is carried out by physical therapists (see page 129).

Strapping for club foot

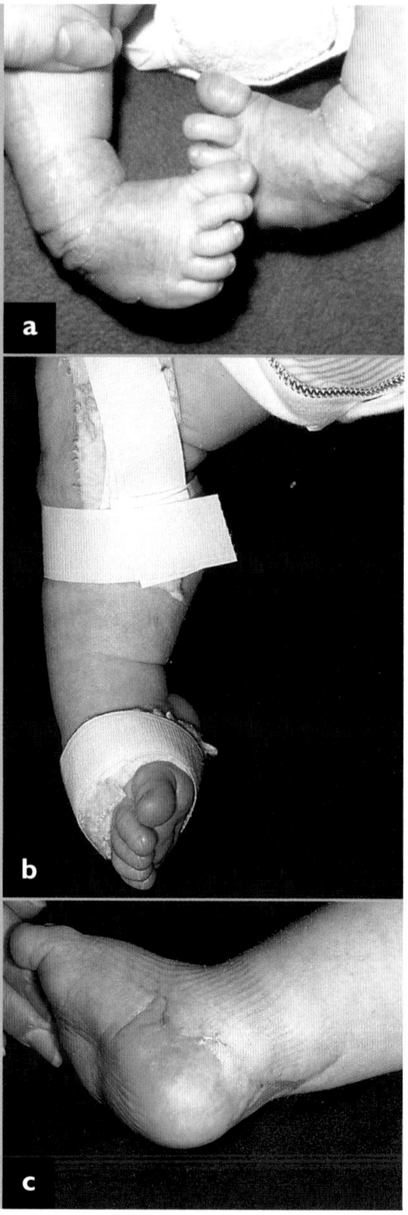
a
b
c

a This newborn baby has the most common type of the congenital deformity known as club foot (talipes equinovarus), in which the feet are twisted downward and inward, with the heel shortened and retracted. The condition is more common in boys than girls, and about half of all cases involve both feet.

b The soft bones of babies will move if constant pressure is applied. Initial treatment of club foot seeks to take advantage of this malleability of a baby's bones and soft tissues by applying adhesive strapping (as here), splints, or a cast to the lower leg and foot to hold the foot in a corrected position. The splints or cast are changed every few days to gently force the foot, little by little, into a more normal position, with the sole facing the ground.

c If this fails to correct the deformity, surgery is performed to release the constraining soft tissues (ligaments and tendons), and the use of splints or a cast is then continued. Shown here is the foot of a seven-month-old baby after the successful treatment of a club foot by means of an operation and splinting.

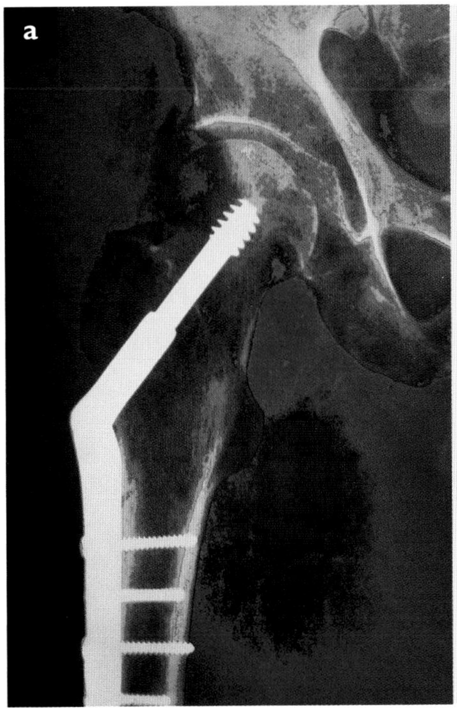
a

Internal and external surgical fixation

a A fractured femur (thigh bone) is held together while it heals by a long plate fixed to the bone with metal screws. At the top of the plate is a nail that slides within its holder to allow the fracture to compact a little when the patient places weight on the leg.

b A fractured tibia (shinbone) is fixed in place after realignment by metal pins inserted through the skin and into the bone. These are held in place by an external metal frame. Once the tibia is secured, there is no need to pin the fibula as well.

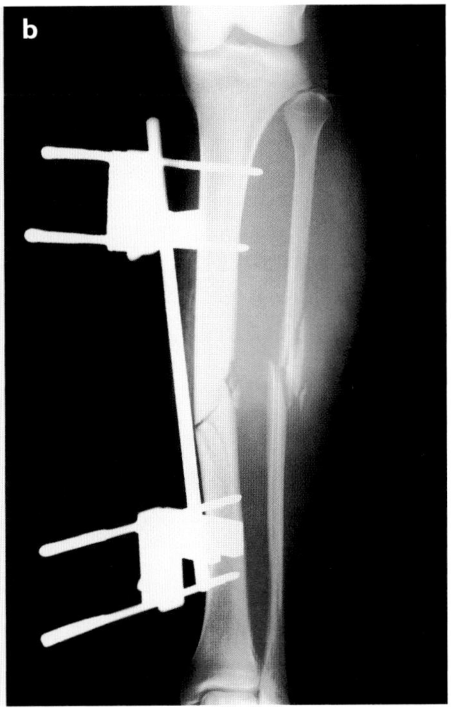
b

FIXATION DEVICES THAT REQUIRE SURGERY

The other class of devices that give support to injured bones and joints requires surgery to put the devices in place. The main advantages of surgical implant devices are that they can control smaller bone fragments and that they allow the patient greater mobility earlier in cases of fractures of long bones. Internal fixation generally eliminates the need for the patient to wear an additional cast or splint, but sometimes this is required.

The obvious drawback is the need for surgery with its ensuing complications—especially the risk of infection, which is increased by the presence of foreign material in the body. The implants are usually made of metal (commonly stainless steel or titanium): They must not be toxic to the body and must be strong enough to withstand the stresses placed upon them until the affected bone or bones have healed properly.

Depending on individual circumstances, internal fixation devices are sometimes removed after they are no longer needed and sometimes are left in place permanently.

Choices for internal fixation

- Plates and screws are probably the most frequently used devices to fix fractures. There are many different designs, sizes, and combinations, depending on the nature and location of the problem—frequently a fracture of a long bone in the arm or leg.
- An intramedullary nail is a rodlike nail that is inserted at one end of the bone or at the fracture site and is positioned lengthways down the middle of a fractured bone. This method of fixation often allows early mobilization of a limb and even some degree of weight bearing. The biggest disadvantage is that if an infection develops, it can quickly spread through the whole bone.
- Wires are sometimes used to hold small bone fragments together and in place. Alternatively, a single screw is sometimes used.

External frame fixation

External fixators are based on pins placed or screwed into the bones, above and below the fracture, and connected to a frame that is outside the body. Their design varies according to the nature of the injury and the part of the body concerned. There are external fixators designed for sites as diverse as the hand and the pelvis. Advantages of external fixators include

- the possibility of correction of deformities after the operation by adjustment of the frame;
- easier access to wounds that need periodic treatment than would be case with a cast or most splints;
- that the fixation metalwork does not interfere directly with the site of fracture; and
- that if infections develop, they tend to be limited to the site of a pin, which usually makes treatment easier.

Disadvantages include the fact that the pins can be uncomfortable, even painful. Also, they are easy entry points for infection. Pins and frame are removed once they are no longer required.

External fixation for bone loss and limb lengthening

The Ilizarov fixator frame, developed in the USSR in the 1960s by Gavril Ilizarov, is used to treat bone loss in severe fractures and to correct some deformities. It can elongate the bones of the lower leg when a patient has one leg shorter than the other or wants to become taller.

Wires held rigid by a frame made up of rings and rods are passed through the bone above and below the fracture point. The rings are then pulled apart very slowly, over many weeks, to increase the gap between the fractured bone ends and so encourage new bone to form in the new space. The technique is based on Ilizarov's observation that the periosteum, the membrane covering the bone, is capable of producing new bone. If the bone is cut, while the periosteum is preserved, and then slowly pulled apart, the periosteum and the broken edges of the bone are stimulated to produce bony tissue.

The technique was introduced to the West in the early 1980s and enthusiastically received before the extent and potential severity of the complications were fully realized. Nonetheless it remains one of the most useful methods of treatment for complex orthopedic problems.

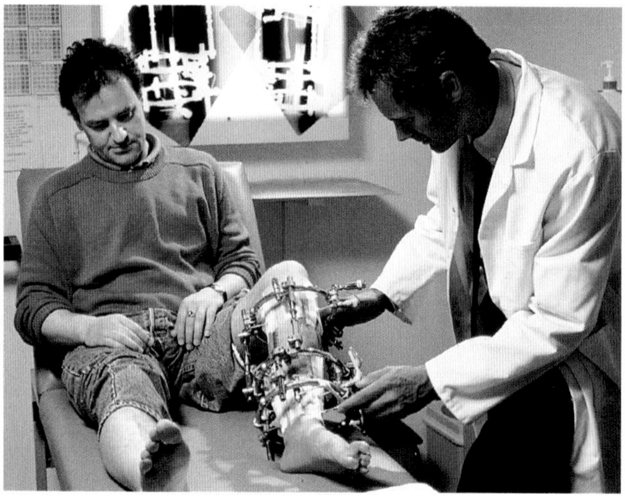

An Ilizarov fixator frame

Complications include a relatively high rate of wire tract infection, chronic pain, and the necessity for frequent assessment and correction. The patient must be prepared to clean the frame's wires on a regular basis and to make tiny adjustments to the frame several times a day.

Amputation and replantation

Fewer amputations take place today because of advances in microsurgery, allowing for the repair of blood vessels and nerves, and even for the reattachment of severed limbs.

AMPUTATION

Nowadays, amputations normally only become necessary when the circulation within a limb (usually a leg) has deteriorated to such an extent that the limb has become gangrenous. Sometimes a surgeon must amputate a body part crushed in an accident (or tidy up a limb or digit already severed). Very occasionally, amputation is needed as the result of a birth defect or to remove a tumor.

For a limb, the ideal point of amputation is far enough below a major joint to keep the full use of that joint. The surgeon shapes the stump to fit a prosthesis (artificial limb) if at all possible. A flap of healthy skin is retained to cover the stump and then stitched in place. For a finger or toe, the best point of amputation is at a joint.

REATTACHING A SEVERED DIGIT OR LIMB

- **Replantation** About 30 years ago, improvements in microsurgical techniques began to make possible the successful reattachment of severed hands as well as fingers, with patients regaining significant sensation and use. The challenge is that blood vessels, nerves, bones, muscles, and tendons must all be rejoined; this is now often possible, as long as the hand was cleanly severed.
- **Transplantation** The attachment of a hand or arm from a (living but brain-dead) donor has only become possible in the last few years, since immunosuppressant drugs have become able to cope with rejection of the replacement by the patient's immune system. There have been a number of successful cases since the first successful hand transplant, in 1998. Five years later, in 2003, Franz Janmig from Austria became the first man to have both arms transplanted. Surgeons have concentrated on transplants involving the hand because it is only here that benefits to a patient outweigh the disadvantages of the long-term health risks from a lifetime use of immunosuppressant drugs and the high cost of an operation that is not needed to save a life.

Surgery for damaged joints

There have been dramatic improvements and developments in orthopedic surgery over recent decades, thanks to a combination of better instrumentation and newer prostheses and devices such as plates and screws.

ARTHROSCOPY FOR A PROBLEM JOINT

An arthroscopy is a surgical procedure employing an instrument called an arthroscope, which is passed into a joint in order to see and diagnose a problem and very often to perform corrective surgery as well. Arthroscopic procedures are most often performed on knee and shoulder joints, but arthroscopy of the elbow, wrist, hip, and ankle is becoming more and more common. Most "keyhole" arthroscopies demonstrate similar levels of proficiency and success as conventional open surgery but with quicker postoperative recovery times. Almost all can be performed as day cases. In certain situations, however, the arthroscope may not offer enough maneuverability and open surgery may be needed as well, either at the time or at a later date.

What happens during an arthroscopic operation?

Arthroscopic surgery normally requires two or more small incisions strategically placed around the joint to avoid injury to neighboring nerves and vessels while still allowing for good visualization and manipulation of the joint. The joint is distended with saline injected under pressure; this opens up space for the instruments while also protecting the joint with a "liquid mattress."

The arthroscope is inserted into the joint through the first incision. The arthroscope is about the thickness of a pen and incorporates an optical telescope that is usually connected to a video camera to display the interior of the joint on a monitor. Instruments such as scissors and forceps are passed into the joint through another incision. The surgeon uses a small metallic hook to probe the joint and test ligaments. After this, any excess synovial membrane is cleared with a rotating blade and sucked out of the joint by a pump. If necessary, the synovial membrane can be sent for pathological examination. Stitching and reshaping of bone can then proceed as necessary, pieces of torn cartilage or fragments of broken bone may be removed, and ligaments can be reconstructed.

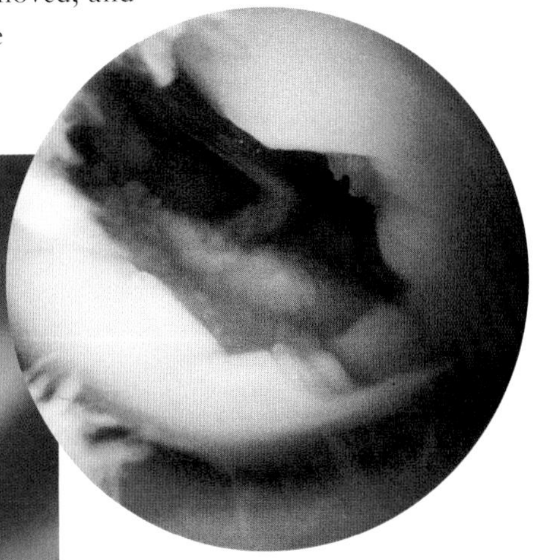

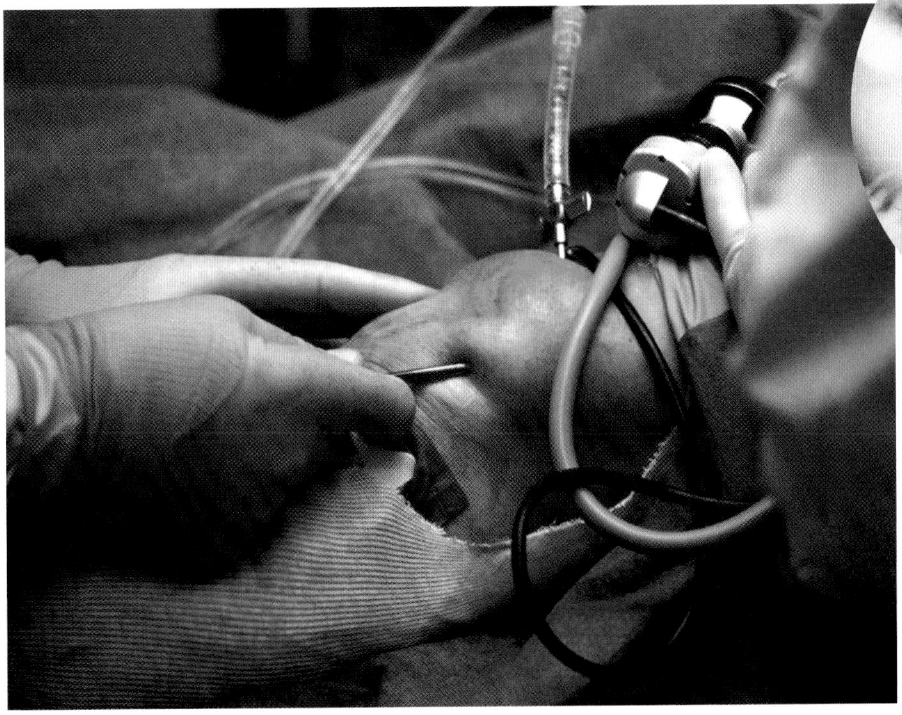

Arthroscopy of a knee joint

An arthroscope is a type of endoscope that is inserted through a narrow incision in the skin to allow a surgeon to view and then reshape or remove part of the knee joint. Left: a surgeon inserting an arthroscope into a patient's knee. The knee has been painted with antiseptic iodine. Above: an arthroscopic view of tissue death of the femur at the knee.

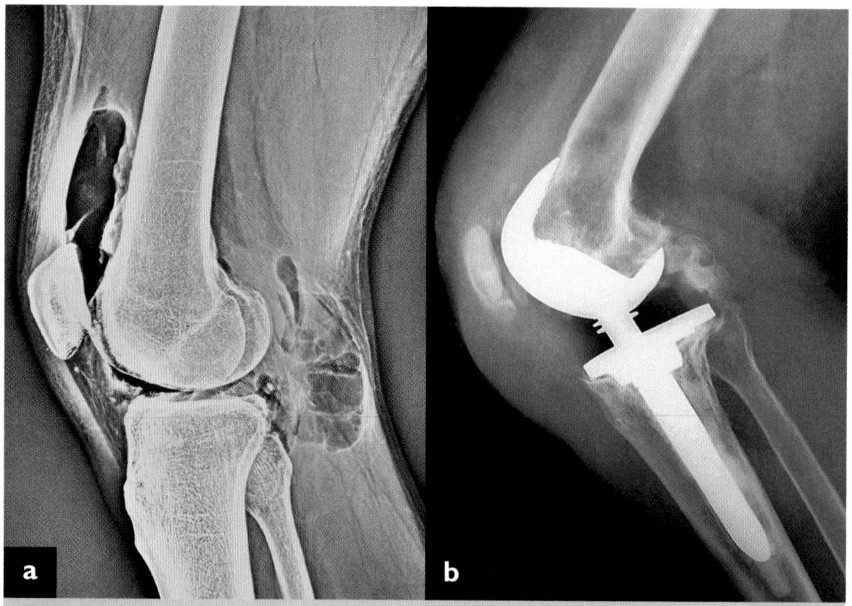
a
b

Knee joint replacement for osteoarthritis

a A knee with osteoarthritis—the disease causes wasting of the cartilage and thus increasing friction between the bones of the joint.

b The implant forms a flexible joint that can hinge like the real joint. The surgeon opens the knee joint and removes any debris, then shapes the bones for the implant. The femur above the knee is cut and the femoral part of the prosthesis is fixed in place. A component consisting of a "tray" on which the femur rests and a tapered post inserted in order to secure the tray in position are fixed into the tibia. The parts might be cemented into place or they may have meshlike edges through which the bone can grow and fuse to hold the prosthesis securely.

JOINT REPLACEMENTS

The goals of surgically remodeling a joint by replacing one or both bone ends with a metal or plastic prosthesis (known as replacement arthroplasty) are pain relief and restoration of a functional range of movement. The most common joint replacements are those of the knee and the hip (see pages 122–23). Of the rest, the most performed are replacements for the shoulder, the elbow, and the joints of the hand, particularly between the metacarpals and phalanges. The wrist and the ankle have proven difficult to substitute and so joint fusion remains the treatment of choice for osteoarthritis in these joints.

The most serious potential complication after joint replacement is infection, which occurs in 1 to 2 percent of cases and usually means that the prosthesis has to be removed. Also, up to half of all patients may have some degree of thromboembolism (blood clots in the veins, which may migrate to the lungs); the risk is lower in the case of joint replacements in upper limbs.

FUSING A JOINT

The surgical fusion of bones across a joint space (known as arthrodesis) eliminates movement and so is a drastic course of action but is sometimes necessary for a joint that is particularly painful or unstable and for which, for some reason, a joint replacement is not the best option. The most common way of fusing a joint is to strip the cartilage from the surfaces of the bones that come into contact with each other and then to bring the exposed bony surfaces together with a cast or internal fixation, thereby immobilizing the joint until the bones have naturally fused together.

SURGERY FOR LIGAMENT AND CARTILAGE INJURIES

Ligaments are the fibrous bands that hold bones in place at a joint. Tearing a ligament is a common sports injury (see pages 143 and 153). For complete ligament rupture, surgical repair is sometimes the answer. The divided ends are stitched together or the ligament is reconstructed with a graft and the affected joint is immobilized with a cast or a splint or brace while the ligament heals. Torn cartilage that causes increased friction between the bones of a joint—typically the knee—is sometimes repaired or trimmed surgically.

SURGERY FOR A TENDON RUPTURE

A tendon is the band of fibrous tissue that connects a muscle to a bone. When treating a ruptured tendon, surgery may or may not be the answer (see page 151). Stitching together the divided tendon is sometimes enough, followed by immobilization with a cast or splint until the tendon has healed. Sometimes a surgeon will repair a tendon using a graft taken from a tendon elsewhere in the body or from a cadaver donor; this is called a tendon graft.

JOINTS SUITABLE FOR REPLACEMENT

Nine out of 10 patients who have knee, hip, shoulder, or hand joints replaced do not need additional surgery for 10 years or more. Elbow replacements have a less successful track record but are still performed because the handicap from a fused elbow is so great.

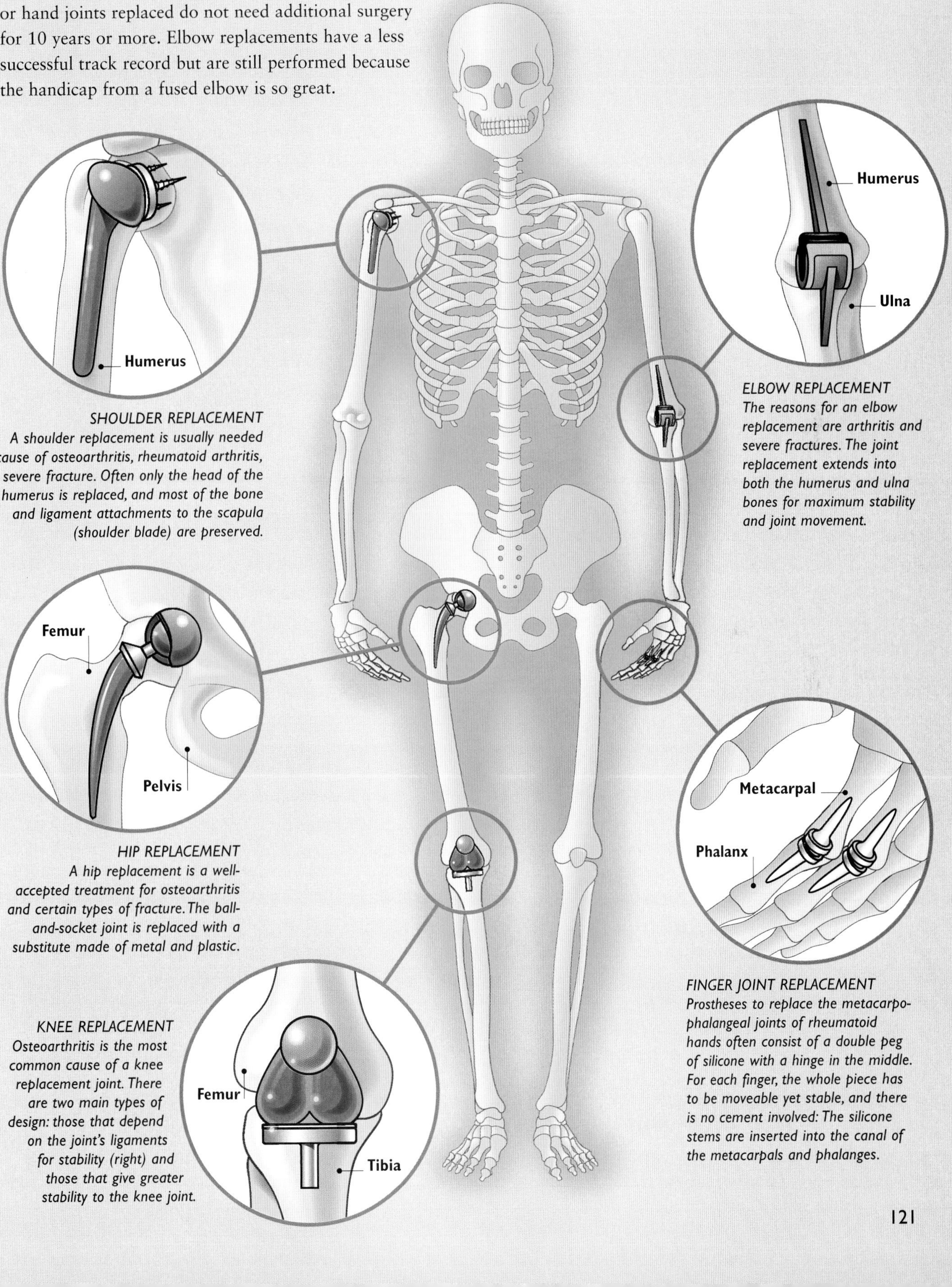
Humerus
Humerus
Ulna
Femur
Pelvis
Metacarpal
Phalanx
Femur
Tibia

SHOULDER REPLACEMENT
A shoulder replacement is usually needed because of osteoarthritis, rheumatoid arthritis, or severe fracture. Often only the head of the humerus is replaced, and most of the bone and ligament attachments to the scapula (shoulder blade) are preserved.

ELBOW REPLACEMENT
The reasons for an elbow replacement are arthritis and severe fractures. The joint replacement extends into both the humerus and ulna bones for maximum stability and joint movement.

HIP REPLACEMENT
A hip replacement is a well-accepted treatment for osteoarthritis and certain types of fracture. The ball-and-socket joint is replaced with a substitute made of metal and plastic.

KNEE REPLACEMENT
Osteoarthritis is the most common cause of a knee replacement joint. There are two main types of design: those that depend on the joint's ligaments for stability (right) and those that give greater stability to the knee joint.

FINGER JOINT REPLACEMENT
Prostheses to replace the metacarpo-phalangeal joints of rheumatoid hands often consist of a double peg of silicone with a hinge in the middle. For each finger, the whole piece has to be moveable yet stable, and there is no cement involved: The silicone stems are inserted into the canal of the metacarpals and phalanges.

Hip replacement

The key factors in a successful hip joint replacement are an implant made of materials that combine high durability with low friction, careful resurfacing of the bone that will touch the implant, and effective fixation of the implant to the bone.

Hip replacement became generally available in the 1970s. Today, it is a well-accepted form of treatment for an osteoarthritic hip and certain types of fractures. The procedure involves replacing the head and neck of the femur with an artificial joint. The head of this implant is ceramic or cobalt chrome, and the shaft is different materials. The implant's stem is attached to the femur using acrylic cement. The hip cavity commonly has a metal backing with a polyethylene center.

When a joint is implanted in the body, the wear process starts immediately. Unlike their biological counterparts, implants are not renewed or repaired by the body and so may wear down or loosen in the bone. Patient selection is key to the long-term success of the replacement. In the past, most surgeons would not operate on patients younger than 65 (because a joint might need to be replaced again) and heavier than 180 pounds. With the steady improvement in component design and techniques for fixing the implants to bone, joint replacements are now available for younger and heavier patients.

An ongoing story

Eight out of 10 hip joint replacements are still in place 20 years after implantation without additional surgery having taken place. If part of the implant fails, it is generally possible to replace it without renewing the entire implant. This is good news, because with each successive operation more bone is lost from around the implant. Eventually, bone grafts become necessary and each procedure is accordingly more technically demanding, with higher risk for the patient and a lower return in terms of level of activity and pain relief.

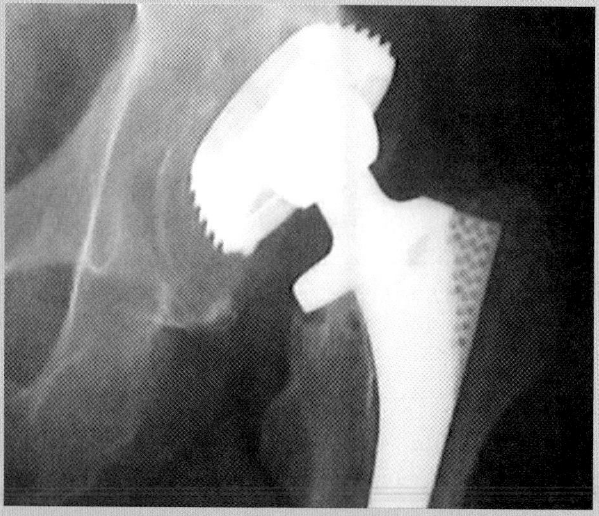

Replacement parts
This X ray shows that the metallic cup in the pelvis is shifting out of the bone (upper right corner) and needs to be replaced. The stem in the femur is still relatively well sited.

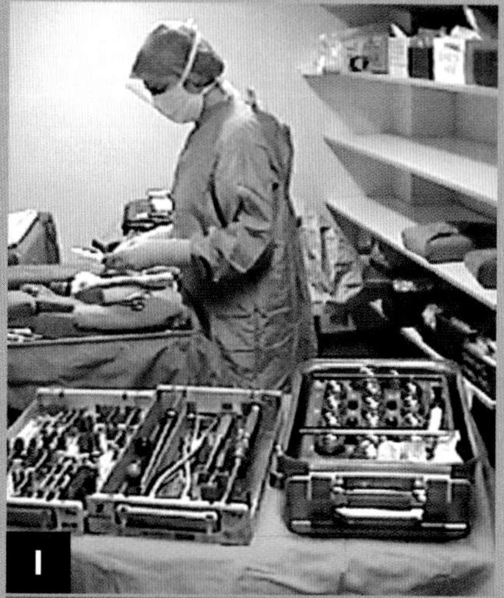
1

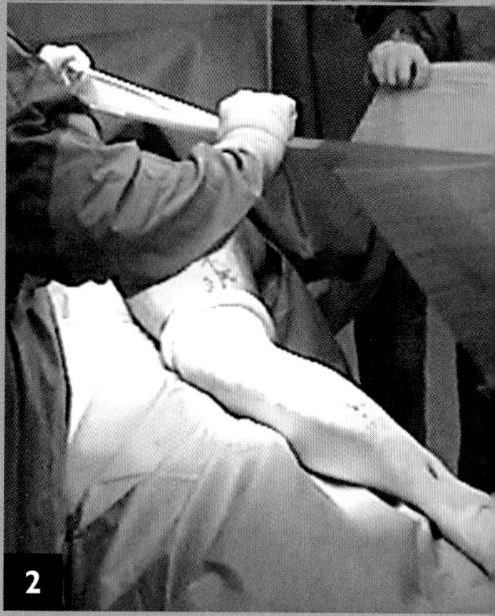
2

The procedure—step by step

1 The scrub nurse checks all the components and has them organized before the operation begins. She is also responsible for making sure that the surgeon gets all the instruments at the right time during the operation and for counting all instruments and swabs to be sure none are left inside the patient.

2 Once the patient is anesthetized and positioned on the operating table, the hip to be operated on is isolated with sterile, waterproof fabrics, and an adhesive plastic sheath impregnated with iodine is used to cover the skin. It is the iodine that gives the yellow tone to the skin and the plastic sheath.

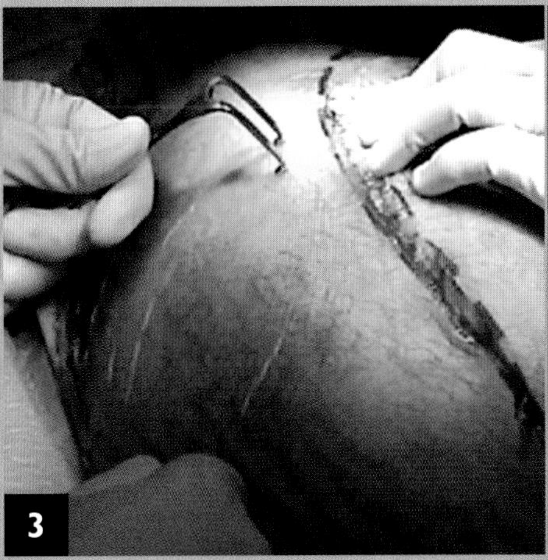

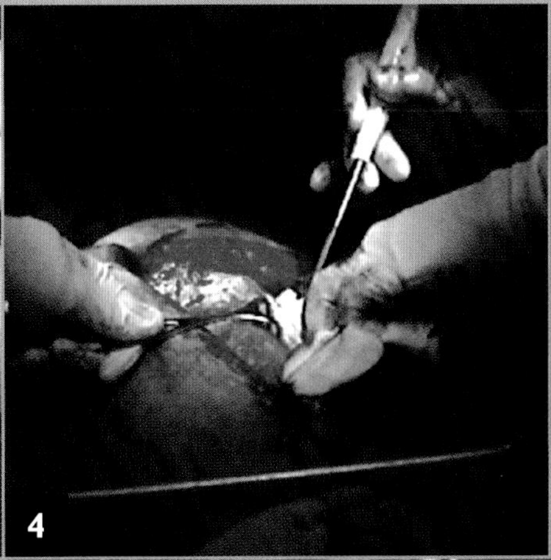

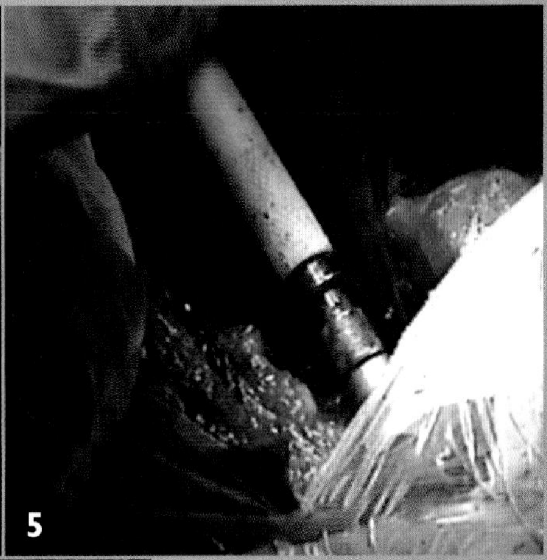

3 The surgical staff wears waterproof sterile gowns, gloves, and masks to reduce the risk of infection. The anesthetist also gives intravenous antibiotics to the patient. The skin is marked with a black pen to show the line of the incision and then is cut using a small, sharp knife.

4 To control bleeding, most surgeons use an electric cautery device: This is a metallic stem that generates high temperatures in a small area, allowing the surgeon to burn and seal the tips of bleeding vessels.

5 An instrument called a reamer is introduced into the hip socket. This is used to remove remnants of dead bone to prepare the bed for the pelvic component of the prosthesis.

6 A similar procedure is performed on the femoral component, removing the top of the femur and reaming the canal. Trial prostheses are used to test the fit of the components. The final choice is shown here before insertion (inset) and in place, in contact with the pelvic component. Pressure between the two components is necessary so they do not dislocate.

7 The wound is washed thoroughly with several liters of saline, before and after the insertion of the implant. The white gun delivers saline under pressure through one channel while a separate tube aspirates the used liquid.

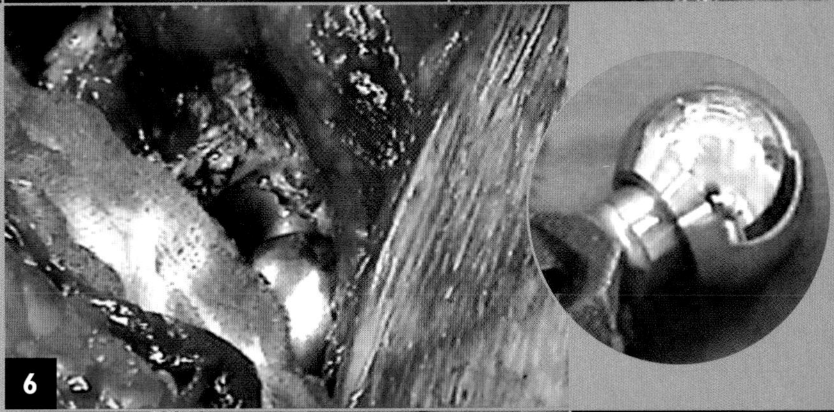

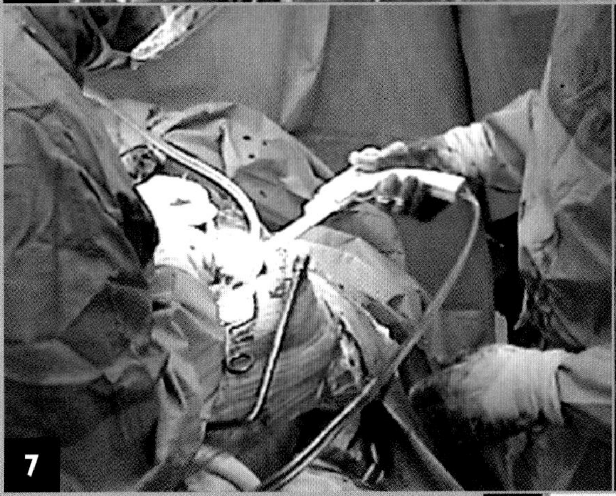

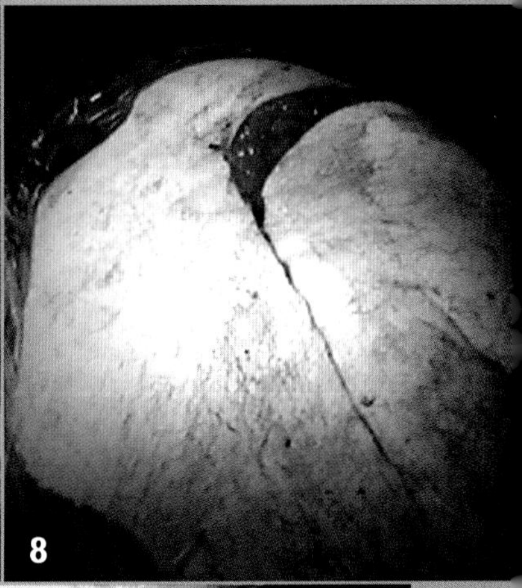

8 The layers of tissue are repaired. Usually one layer of stitches is applied to the deep tissues such as the ligaments and another, if necessary, to the muscles (the muscles are often retracted and not cut). Finally the skin is stitched, in this case with absorbable (dissolving) stitches that will not need removal later.

9 A colored X ray shows the prosthesis in place.

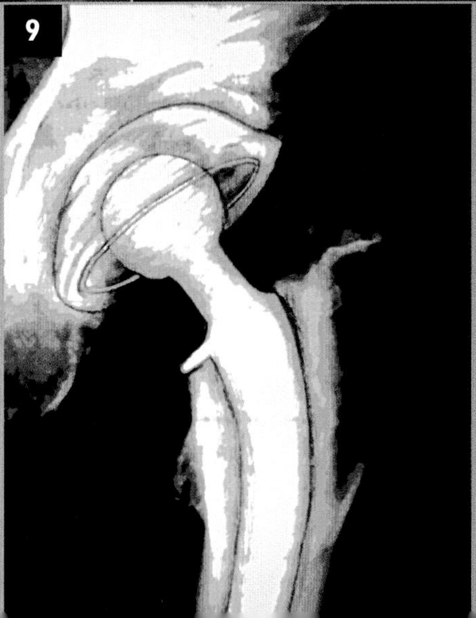

Surgery on the spine

Most operations performed on the spine are carried out to alleviate problems and pain caused by wear and tear, or injury. Plus, there are operations performed to rectify problems patients are born with or develop as they grow up, such as curvature of the spine (scoliosis).

Endoscopic and microsurgical techniques are making many back operations less traumatic than they used to be, but spinal surgery is still often the last resort, after other treatments have failed, because some trauma remains and complete success cannot be guaranteed.

SURGERY FOR A BROKEN BACK

A patient is said to have a "broken back" when one or more vertebrae are fractured. The severity of the problem, and the type of treatment (surgical or otherwise) vary greatly from case to case, although the basic principles of support and immobilization of the affected area apply. The final outcome for patients varies from complete recovery to paralysis depending on the injury.

DISCECTOMY

A discectomy is a surgical procedure in which all or part of an intervertebral disc is removed. Discectomies are performed for a number of reasons, but the most common is to relieve pain from a prolapsed disc.
The operation is done either by using traditional open surgery or, increasingly, minimally invasive microsurgery.

A prolapsed "slipped" disc

This CT scan of a prolapsed disc shows the intervertebral discs as yellow, the bony vertebrae as pale blue, and the spinal cord as dark blue. The outer layer of the prolapsed disc has ruptured, and its pulpy contents are bulging out and pressing against the nerves of the spinal cord, causing pain and numbness. If other treatments fail, surgery may be the answer—generally a discectomy. For more information on prolapsed discs see page 148.

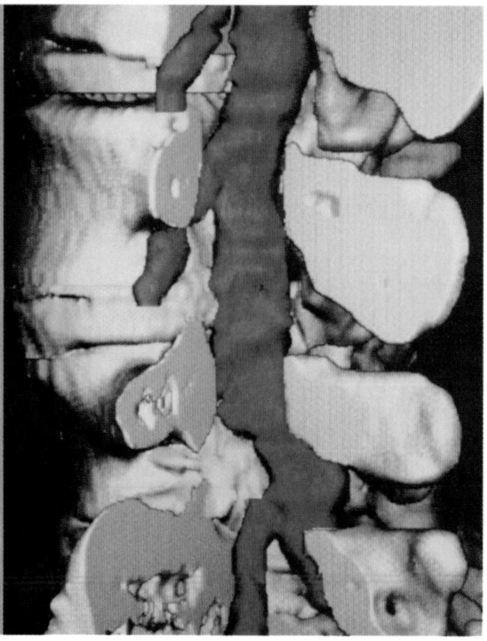

A discectomy for a prolapsed (slipped) disc

The prolapsed (or slipped or herniated) disc is partially removed in order to relieve pressure and irritation on the nerves within the spinal cord. As these intervertebral discs—made of fibrocartilage—are needed to act as buffers between the vertebrae, the surgeon tries to remove only the fragment of the disc that is protruding into the spinal cord, leaving as much disc tissue in place as possible.

When performing a microdiscectomy, the surgeon is guided by the view through a microscope. Disc material is removed bit by bit through a small incision, until the nerves of the spinal cord are no longer compressed. The surgeon can operate very precisely, with minimal damage to the surrounding tissues, and so there is a faster recovery time for the patient after the operation than would be possible with open surgery.

SPINAL FUSION

In bone fusion surgery, two bone surfaces are brought together, and as they heal, they grow together and become fused. When this is done in the spine, the aim is to create greater stability between two or more vertebrae; the drawback can be reduced flexibility.

- **Long fusions** across a number of vertebrae are most often performed to correct cases of curvature of the spine (see scoliosis, page 151, and kyphosis, page 142).
- **Short fusions** of two or three vertebrae are carried out to stabilize a range of infirmities brought about by deterioration caused by aging or accident.

There are different techniques for bringing two bone surfaces together so that fusion can take place.

- **A bone graft** Often, a surgeon will bring in bone from elsewhere to act as a bone bridge, or wedge, between two bones that need to be fused. The imported bone (the

EXPERIENCING A MICRODISCECTOMY

About a year ago, I started to experience stiffness and discomfort in my lower back and pain in my right leg. An MRI scan revealed the problem to be a prolapsed disc in the lower back. Physical therapy and drug treatments did not improve my condition, so the consultant recommended that I have a microdiscectomy.

I was in the operating room for a little over an hour. I awoke to find myself lying on my back, but I was periodically rolled from left to right to relieve pressure on the operation site. After a few hours had passed, I was taught how to maneuver my back without placing undue strain on the wound in my back. Technology has advanced so far that I was able to go home within 24 hours.

I returned to the hospital on the 10th day after the operation to have my stitches removed. The physical therapist had worked out a program of gentle exercises and swimming for me in order to stabilize and strengthen my spine, and I was advised of the importance of keeping to this program for making a full recovery.

I was able to go back to work after four weeks. I still have to be careful and can't carry anything too heavy, but I am in far less discomfort than I was previously.

bone graft) may come from another part of the patient's own body or from a cadaver donor or be synthetic bone. Bone taken from the patient may come from the pelvis, the fibula in the lower leg, or a vertebra close to the operation site. A long fusion may involve a succession of bone grafts along a length of the spine.

- Removal of intervening tissue The tissue that separates the two bone surfaces is removed so that bone can be placed against bone. This may involve a discectomy.

After the bony surfaces have been placed against each other, the affected area is stabilized and immobilized with a brace, cast, or some form of internal fixation. External supports are removed as soon as the bones have fused in place. (This can involve a wait of some months.) In contrast, surgically implanted "internal fixation"—generally consisting of a plate or rod plus screws or a more complex "cage"—usually remains in place permanently, often being very difficult to remove.

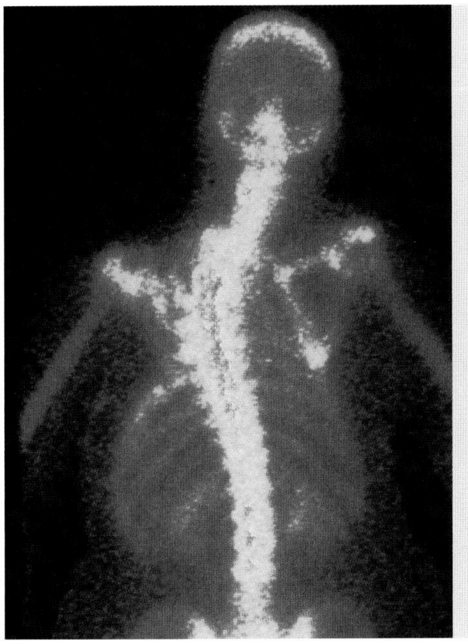

Bone fusion for scoliosis

The word scoliosis means a lateral curve or tilt of the spine (as shown left, seen from behind). This condition may be present at birth or may develop later on. Surgical treatment to correct severe scoliosis varies from case to case, but it often involves the realignment of the spine by means of bone fusion and internal fixation. The corrected position is often fixed in place by a metal rod and screws (right) and wires, which remain in place even after the bones have fully healed.

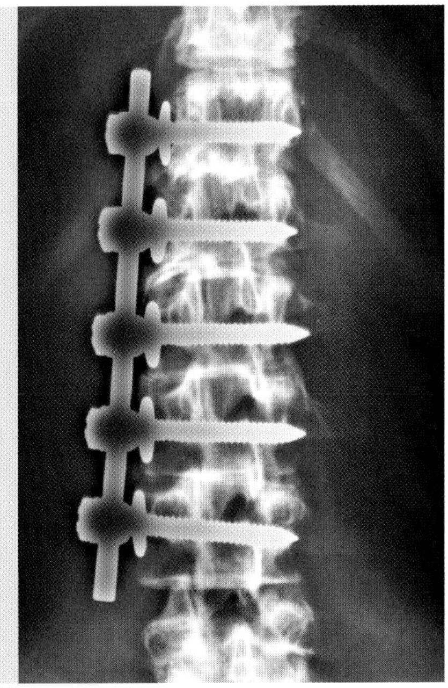

Physical therapy

Physical therapists use physical rather than chemical or surgical methods to treat a wide range of conditions. They play a key role in rehabilitation following accidents, surgery, and long-term illness. They not only treat injuries but also teach patients techniques to help avoid any recurrence.

EXERCISE THERAPY

All physical therapy treatment for musculoskeletal problems include exercises, which need to be performed regularly to be effective. Initially, the physical therapist supervises and teaches the patient, who is often then able to carry on unsupervised at home.

Exercise has four main purposes:

- To mobilize joints to their full range of movement.
- To reduce muscle tension and muscular imbalance.
- To increase muscle tone and strength.
- To improve coordination and balance.

The type of exercise chosen depends on the degree of immobility, but generally there is a progression from passive movement to resisted exercise as the treatment begins to take effect.

- Passive movements The physical therapist manipulates a joint in order to maintain its range of movement. These exercises are used when patients are unable to move the joints themselves because the muscles are too weak or paralyzed or movement is too painful.
- Isometric exercises The isometric contraction of a muscle occurs when a muscle contracts but there is no movement at a joint, such as when you pull your stomach muscles in to flatten your tummy.

These exercises are used when a movement causes pain, or when a joint is immobilized— by a cast, for example.

Hands on and hands off

Many physical therapy techniques used in the treatment of musculoskeletal problems utilize the physical therapist's manual skills alone, but some require specialist equipment.

a A physical therapist manipulates a patient's neck. This is called passive exercise.

b A patient engages in active, resisted exercise using an Orthotron machine to restore leg muscle strength. The machine allows the leg to move in one direction only and exerts a measured force against the direction of movement.

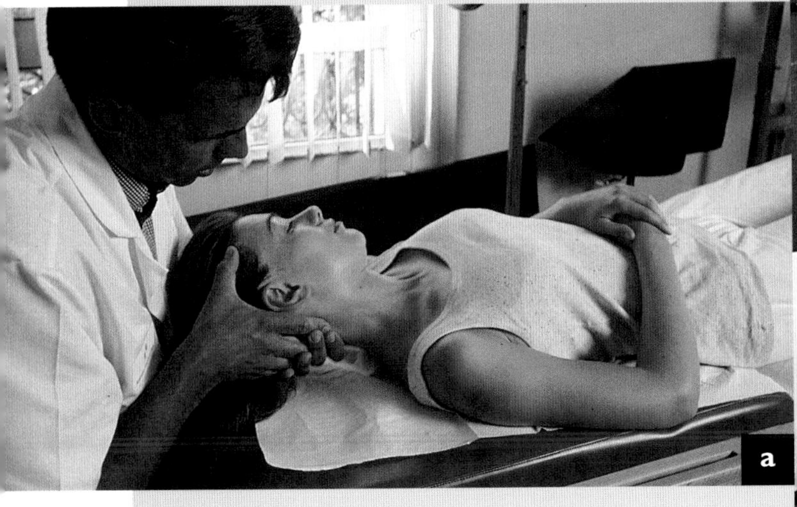

a

b

- **Active exercises** The patient moves a joint under the supervision of the physical therapist. Active exercises are used to reeducate and strengthen muscles, to correct any muscular imbalance, and to increase the range of movement and flexibility at a joint.
- **Resisted exercises** The patient moves against an opposing force, either from the physical therapist or weights. Again, these are used for strengthening the muscles and to increase the stability of a joint.

Physical therapy exercises are often performed in water, in a hydrotherapy pool (see page 130), though not all hospital physical therapy departments have access to one. The warmth and buoyancy of the water reduce tension and help the muscles relax, allowing active, pain-free movement of the joints. This increases the range of joint movement while the exercises build up muscle strength.

MANIPULATION OF JOINTS

Physical therapists often use their hands to gently manipulate the spinal joints to correct spinal alignment. On other joints, such as the shoulder, they sometimes perform "aggressive" manipulations. These are more forceful, larger-scale movements designed to break down a build-up of tissue that has led to loss of elasticity (called a "contracture" when severe and an "adhesion" when less serious) in the fibrous tissue comprising ligaments, joint capsules, and cartilage. They must be performed with care. Adhesions often affect the muscles around a joint after an injury, causing stiffness and pain, but can be broken down by passive and active stretching as well as massage.

MASSAGE OF SOFT TISSUES

The type of massage that physical therapists use is known as "Swedish massage," after Henreich Ling, a Swede, who developed this system for massaging the body's soft tissues for therapeutic effect. There are six main movements in Swedish massage, and individually or in combination, they affect the skin, muscles, tendons, ligaments, joints, blood and lymph vessels, nerves, and some underlying organs. However, they are particularly useful in the treatment of musculoskeletal problems.

- **Effleurage** Used before and after other movements, this involves a stroking action using the flat of the hand. It increases the circulation of both blood and lymph.
- **Petrissage** The balls of the thumbs or fingers are used to squeeze soft tissue against an underlying bone. This reduces tension in muscles and promotes speedier removal, via the bloodstream, of the chemical wastes produced when muscles contract.
- **Friction** This is a deep, circulatory movement, applied using the ball of the thumb. It is used to break down knotted, tense muscle fibers and contractures. Friction can be painful, so it is only used for short periods.
- **Kneading** This is a movement similar to that used when kneading bread. It is applied to large areas of soft tissue, such as the thighs, to reduce tension, improve the ability of muscles to contract, eliminate waste products, and increase the flow of blood to the area.
- **Hacking** The edge of the hand is used in a way similar to a karate chop. It improves muscle tone and reduces tension, especially in the back and shoulders.
- **Cupping** Light, rhythmic blows with cupped hands improve the blood circulation of the skin and the tissues just beneath it.

ELECTRICAL TREATMENTS

The nervous system relays both sensory information, such as the perception of pain, and command instructions, such as those that tell a muscle to contract, by means of

TALKING POINT

The problem with backs

More working days are lost as a result of back problems than for any other reason. Problems may be muscular or skeletal or a combination of the two. They may also be chronic (long-term and progressive) or acute (of sudden onset). These categories can be blurred when a chronic problem suddenly worsens and become acute. The most common cause of back problems is poor posture. Our upright stance puts a tremendous strain on the spine, which also has to absorb shock waves from the ground as we walk. To counter this, the spine has shock-absorbing cushions between each vertebra and natural curves along its length. Most people either hold themselves in an overly upright manner, with taut, tight muscles and flattened curves, or are slouched, with rounded shoulders and overemphasized curves. Correct posture, somewhere between these extremes, ensures that pressure on the spinal joints is even and muscles are loose rather than taut or stretched.

electrical impulses. Exploiting this, physical therapists often use equipment that transmits electrical impulses as part of their treatment programs.

- Transcutaneous electrical nerve stimulation (TENS) A TENS device transmits minute electrical impulses to the nerve endings in the skin to relieve persistent pain. The treatment stimulates the nerve receptors responsible for touch, which relay messages to the brain more quickly than other receptors, and these faster impulses block out the slower pain impulses from the pain site.
- Faradism This is used to help muscle tone after a period of forced inactivity. Pads are placed over the muscle to be treated, then an electrical charge stimulates the nerve that controls the muscle, causing it to contract.
- Galvanism This is the application of a sustained, controlled electrical shock directly to a muscle in order to make it contract. The technique is used when a muscle has no motor nerve supply, usually as a result of an injury, in order to prevent fibrous tissue from building up in the muscle.

HEAT TREATMENT

Heat can be applied to the skin and superficial tissues by means of hot packs, infrared lamps, and wax baths. A short-wave diathermy machine will direct heat that is the product of electrical impulses to the deeper tissues and joints. Heat stimulates and improves the blood supply to the area, eases muscle spasms, and relieves pain. It is used in the treatment of arthritis and of muscle and tendon tears.

CRYOTHERAPY

Physical therapists employ cold packs, sprays, or crushed ice to reduce inflammation, bruising, and swelling following an injury and also to block pain pathways in the nervous system, when necessary.

ULTRASOUND

Physical therapists use touch (and, often, an X ray) to find out precisely where a problem is and how deep in the body it is and then set the ultrasound machine to produce a continuous signal or pulsed wave of high-frequency sound. Pulses help to

Three treatments for the wrist and forearm

Physical therapists have at their disposal a wide range of gadgets that have been developed for the treatment of different disorders.

a Heat treatment
A patient receiving heat treatment from a diathermy machine for a broken bone in her forearm. She has just had a plaster cast removed and is now beginning general exercise. The heat relieves pain and speeds up the natural healing process by increasing blood flow.

b Ultrasound therapy
High-frequency sound waves are used here to treat an injured wrist. Potential benefits include reduced pain and inflammation, increased mobility of the affected joint, and reduced recovery and disability time for the patient.

c TENS treatment
TENS (transcutaneous electrical nerve stimulation) is used here to relieve pain from a damaged radial nerve in the forearm. Pulses of low-voltage electricity are transmitted into the tissues concerned via electrodes attached to the skin.

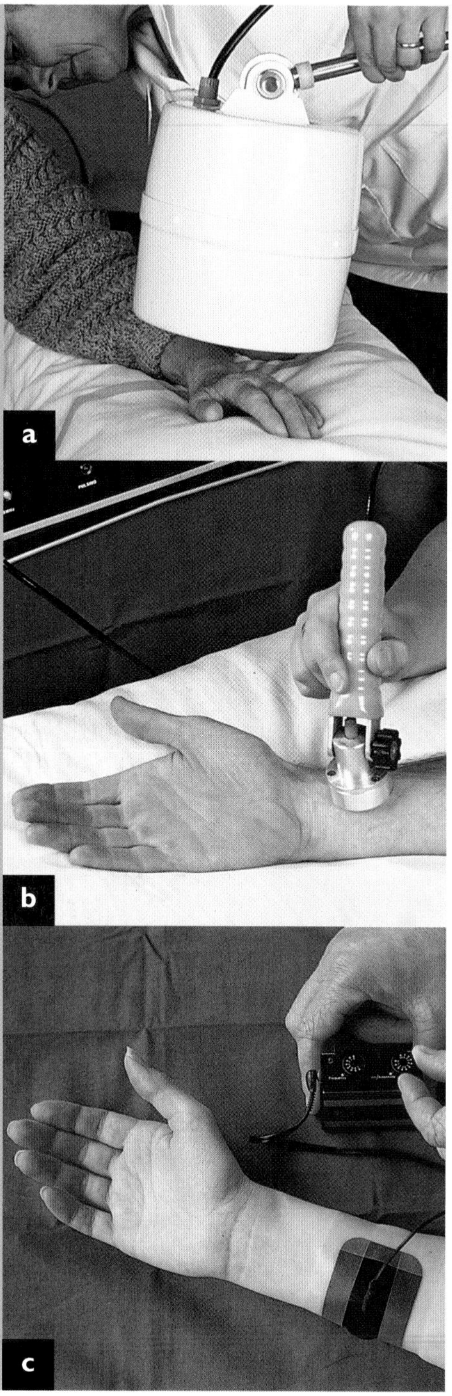
a
b
c

break down fibrous tissue near the skin's surface while a continuous signal is beamed into deeper tissues to reduce swelling, bruising, and adhesions of joints and the surrounding soft tissues. It achieves this by generating heat and by increasing the permeability of individual cell membranes, thus increasing the uptake of nutrients and speeding the elimination of chemical wastes.

LASER TREATMENT

A laser machine produces a concentrated beam of light, which can be of high or low intensity. Physical therapists use low-intensity lasers to treat soft tissue injuries, such as muscle, tendon, and ligament tears, and inflamed joints. The beam not only stimulates tissue healing and reduces swelling, inflammation, and pain by stimulating the flow of blood and lymph but also slows down the production of prostaglandins—the chemicals that cause, among other things, inflammation and pain in damaged tissues.

INTERMITTENT TRACTION

There are two types of traction: continuous and intermittent. Continuous traction is used by orthopedic surgeons, often to realign two parts of a fractured bone. Physical therapists employ intermittent traction to help correct the alignment of bones and to stretch overtight muscles and ligaments. They may apply traction with their hands, or, when more force is needed, use a traction machine. A traction machine is sometimes used, for example, to correct the alignment of the vertebrae, especially in the neck, or to relieve pressure on a nerve caused by a prolapsed spinal disc. The patient lies on a traction table with his or her head in a leather girdle and the pelvis immobilized by straps. Weights are attached to the girdle and the straps by a system of pulleys, encouraging the vertebrae to move apart. Treatment sessions typically last about 15 minutes or so and take place several times a week, but, of course, this varies from case to case.

REHABILITATION AFTER SURGERY

Physical therapy plays a vital role in rehabilitation following any orthopedic surgery, which by its very nature causes musculoskeletal weakness. The aim of treatment is to return the patient to full independence by mobilizing all affected joints, strengthening surrounding muscles, and regaining the flexibility of ligaments and tendons.

If a patient is confined to bed—receiving traction for a fractured femur, for example—isometric exercises maintain muscle strength in the affected area. More general exercises stimulate the circulatory system, maintain overall muscle fitness, and keep bedsores at bay.

Treatment is extended once the patient is mobile, with specific problems each having their own treatment regimen. For example, a ruptured Achilles or patellar tendon that has been mended surgically will be gradually stretched. At first this is done by the physical therapist, using passive movements; then the patient starts to perform active movements, such as using a Wobble Board to increase flexibility and strength in the ankle joint and muscles of the foot. Muscles, tendons, and ligaments may be warmed by heat treatment, massage, or infrared lamps prior to treatment to ease pain, stimulate the circulation, and encourage flexibility. As rehabilitation progresses, the patient moves on to performing resisted exercises, using weights, for example, or an exercise bicycle. If necessary, the physical therapist advises on how to carry out such activities as sitting, standing, walking, and climbing stairs and how to maintain correct posture. Finally, exercises for a home treatment regimen may be prescribed, to continue rehabilitation and prevent the problem from recurring.

EXERCISES FOLLOWING A KNEE JOINT REPLACEMENT

Before the operation to replace the joint in my left knee even took place, the physical therapist drew up a detailed exercise plan and impressed on me the importance of following it to the letter.

On the very first day after the operation, gentle exercise began. I was encouraged to raise and lower my left leg—while it was being supported by the physical therapist—and to contract the muscles in my thigh. On the second day I started to bend the knee. On day three, I began to stand and walk with the help of the physical therapist and a crutch. By the 10th day, I was capable of walking and climbing stairs on my own with the aid of a crutch and so was able to go home.

Outpatient physical therapy and exercises at home then continued until my knee joint was fully functional and all muscle strength had returned.

Hydrotherapy

Many hospitals and clinics offer hydrotherapy treatments because water has three qualities that make it invaluable when treating injuries to muscles and bones: It can be warm or cold, it provides buoyancy, and it offers resistance.

The Romans made water treatments an essential part of their daily lives, building spas that were similar to modern-day health clubs, with baths, treatment rooms, massage areas, and gymnasia.

TEMPERATURE CONTROL

Both warm and cold water can be used to ease muscle tensions and spasms and to relieve aches and pains. Warm water soothes and relaxes by widening the superficial blood vessels, increasing the blood flow to the skin and muscles and reducing the blood flow to the internal organs. Cold water stimulates and invigorates. It also restricts the blood flow to the skin and increases blood flow to the internal organs, thereby inhibiting the biochemical reactions that cause inflammation.

BUOYANCY AND RESISTANCE

Water not only supports the weight of the body but, through its viscosity, it also provides a medium that resists movement slightly. The support reduces strain on weight-bearing joints such as the back, hips, knees, and ankles, allowing for freer and easier movement and mobilization of stiff joints as muscles are being retrained. Meanwhile, the resistance allows for a progressive increase in muscle strength and endurance.

WHO BENEFITS?

Hydrotherapy is an ideal treatment for people with conditions such as rheumatoid arthritis, ankylosing spondylitis, and chronic back problems, which are painful and where it is important to maintain mobility. It is also extremely useful in the rehabilitation of patients

Taking the waters
A patient exercises with weights in a hydrotherapy pool, directed by a physical therapist.

following an orthopedic operation. Hydrotherapy is also used in the treatment of those who are partially paralyzed following an accident, illness, or a stroke, because movement in water feels easier than it does on dry land. This has the psychological effect of enhancing an individual's perception of his or her capability for mobility and so increases self-confidence. Finally, playing in water is fun. For this reason, hydrotherapy is often used to treat children who have painful weight-bearing joints. It can help teach new movements and train breathing abilities.

In the U.S., hydrotherapy is generally used in the treatment and rehabilitation of problems with the musculoskeletal system and of problems such as paralysis, cystic fibrosis, and cerebral palsy. In some other countries, however, such as France and Germany, doctors see hydrotherapy as having an important role in combating stress and promoting general well-being.

Hydrotherapy douching
This is a gentle form of water treatment in which a stream of water is directed at a particular part of the body in order to improve the circulation and relieve pain. Water temperature and pressure are varied according to the needs of the patient.

HYDROTHERAPY TREATMENTS

Today, many physical therapy departments within hospitals have their own hydrotherapy pool, as do numbers of private health clinics.

Exercises in water

In the pool, physical therapists take patients through a range of exercises designed to increase mobility, flexibility, and strength. They often join patients in the pool to keep close control over the specific movements that are made.

High-powered water jets

These direct jets of hot or cold water at problem areas of the body. They are often located in the pool but may also be in a shower room. They can be used to pummel large muscle groups in order to reduce muscle tension and to improve muscle tone and flexibility.

Whirlpool baths

Whirlpool baths—often called Jacuzzis, after the company that pioneered them—are baths fitted with water jets. These jets, positioned along the sides of the bath, propel pressurized bubbles through the water to massage the skin and underlying tissues. The force of the massage can be adjusted to give either a relaxing, soothing sensation or, with increased pressure, an invigorating massage in which the muscles are pummelled and probed for about 20 minutes. The effect is to ease muscular tension, increase muscular and joint flexibility, relieve any pain, and improve the circulation in the skin and superficial areas, thus reducing any tendency for them to retain fluid. The jets in whirlpool baths are useful in the treatment of all types of arthritis, back problems, and general muscular aches and pains, but they are not as precise in their effect as high-powered jets, so whirlpool baths tend to be found in private clinics rather than in hospitals.

Flotation tank therapy

The flotation tank was invented in 1954 by an American neurophysiologist, Dr. John C. Lilly. In recent years, flotation therapy has become more popular and more widely available though, as yet, there are only a few flotation tanks outside private clinics.

A flotation tank is approximately 8 feet long and 4 feet wide and is filled with a dense solution of Epsom salts dissolved in warm water of about skin temperature. The patient simply floats in the water for the duration of the treatment session. The tank is completely enclosed and soundproof, although fresh air circulates through it and an intercom allows the patient to communicate with the therapist. A session lasts for up to two hours and induces a deep sense of calm and relaxation in the patient. It also relaxes tense muscles and reduces chronic pain, so the therapy can be helpful in the treatment of many problems and disorders connected with the musculoskeletal system, such as arthritis, muscle strain, and chronic lower back pain.

Osteopathy and chiropractic therapy

Osteopathy is a system of physical manipulation for the treatment of a wide range of complaints involving the musculoskeletal system without the use of medicines or surgery. Chiropractic therapy takes a similar approach, but treatment focuses on the spinal column.

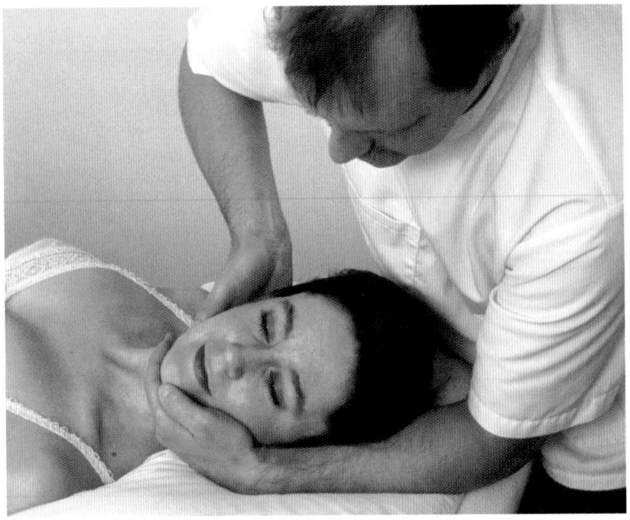

Osteopathy in action

An osteopath using a rotational thrust technique to treat a patient's neck. Other hands-on techniques may include stretching, stroking, and gentle massage of the affected area.

OSTEOPATHY

Osteopathy is a form of therapy founded by Andrew Taylor Still in the United States in the late 19th century. He developed a distinctive "hands-on" approach to physical therapy based on the concept that the musculoskeletal system—bones, muscles, joints, and related tissues, plus the nervous system—plays a central role in the health of the entire body. In fact, osteopathy operates on the principle that the body is a single biological unit with all structures and functions inseparably linked to one another. From this, it follows that local changes can bring about distant changes; for example, a problem with a foot could easily cause additional pain in the back.

Doctors of osteopathy (D.O.s) are one of the fastest growing groups of health care professionals. Between 1989 and 1994, the growth rate of D.O.s in the United States was almost double that of M.D.s.

What disorders do osteopaths treat?

A doctor of osteopathy (D.O.) will treat disorders of the muscles, joints, tendons, and ligaments. Many patients have physical injuries or postural problems, but osteopathic treatments can also help with such varied ailments as headaches, breathing problems, digestive disorders, and even glue ear (fluid in the ear) in children.

Visiting an osteopath

When you see a D.O. for the first time, a personal history will be taken, followed by a visual and manual examination. The D.O. uses a highly developed sense of touch (palpation) to identify points of weakness and problem areas in or around an injury. The D.O. will rarely prescribe medicines or invasive surgery, although he or she may make use of X rays, blood tests, and any other information that the patient can give.

D.O.s use a wide variety of manual techniques, and a session may include one or more of the following:

- Manipulation of a joint, which may range from taking the joint through its normal range of movements to the use of high-velocity thrust techniques.
- Massage of soft tissues.
- Stretching and manual traction where required.
- Pumping techniques to improve blood circulation and drainage of the lymphatic system.

A treatment session usually lasts about half an hour. Different problems require varying numbers of treatment sessions, the average number of visits being six to eight. D.O.s (and chiropractors) also advise patients on aftercare and exercise outside treatment sessions.

CHIROPRACTIC

Chiropractic therapy also originated in the United States; it was developed by David Daniel Palmer at the beginning of the 20th century. "Chiropractic" means "done by hand." In many ways, chiropractic therapy resembles osteopathy, but there is a greater emphasis on the spinal column. Chiropractors take account of all the structures of the body, including the musculoskeletal, neurological, vascular, nutritional, and emotional environments, but they believe that many common problems can be linked to the effects of misalignment of the spine on the central nervous system. Therefore, they place great emphasis on freeing up and mobilizing the spinal column.

A TO Z

OF DISEASES AND DISORDERS

This section gives information on the main illnesses and medical conditions that can affect the bones and muscles.

This index is arranged alphabetically, and each entry is structured in a similar way:

What are the causes?

What are the symptoms?

How is it diagnosed?

What are the treatment options?

What is the outlook?

ANKYLOSING SPONDYLITIS

An arthritic condition affecting young adults in which there is inflammation of joints in the back.

Young men are affected more often than women. The incidence of the disease varies: It is common in certain groups of native North Americans and uncommon in African people.

What are the causes?

The cause is unknown, but the disease often runs in families, so inheritance clearly plays a part. People with the human leukocyte antigen HLA-B27, a particular protein on the membrane of their body cells, are more likely to suffer than those without it (see below). However, the antigen may also be found in healthy people and in individuals with certain other diseases.

What are the symptoms?

The main symptoms tend to be pain and stiffness in the buttocks and lower back caused by inflammation in the sacroiliac joints, which connect the vertebral column and the hip bones. These symptoms often occur in episodes, are usually worse first thing in the morning, and can be relieved by moving around. Pain in the chest is another possible feature, resulting from inflammation of the rib cage, and other joints, particularly the hips, may also become inflamed. Without appropriate treatment, curvature of the spine may develop, together with weakening of the muscles of the back. Eventually, the neck becomes stiff.

Other possible features include inflammation of the aorta (the large artery that carries blood rich in oxygen away from the heart), and uveitis, which is inflammation of the eye. In acute cases, uveitis requires urgent treatment: If a person has eye pain or blurred vision or is bothered by bright light, urgent medical attention is needed.

How is it diagnosed?

Blood tests, such as an ESR and CRP (p. 109), may indicate that inflammation is present somewhere in the body, but they cannot confirm the diagnosis. Testing for HLA-B27 provides another clue, but not everyone with the disease has it. X rays of the spine and sacroiliac joints may show signs of ankylosing spondylitis.

What are the treatment options?

Morning exercises are essential to maintain the spine's mobility and prevent it from becoming deformed. Nonsteroidal antiinflammatory drugs are taken during flare-ups. Steroid injections into inflamed joints may be helpful.

What is the outlook?

The disease has a good prognosis, provided the recommended exercise program is started early and painkillers and antiinflammatories are taken when necessary. Back stiffness may be a problem but is unlikely to cause disability.

What are human leukocyte antigens (HLAs)?

HLAs are proteins present on the cell membranes of nearly all the cells of the body. Also known as major histocompatibility proteins, they are recognized by the body's immune system, enabling it to determine whether the cells it encounters are body cells that should be there or foreign cells, such as cancer or infective organisms, which it should attack.

There are many different human leukocyte antigens. Every individual has a small, specific set inherited from his or her parents. Identifying this set is an important part of finding a suitable match for an organ transplant. The more similar the set of histocompatibility antigens, the better the match.

Some HLAs are of special interest because they are found particularly in people with certain autoimmune diseases. HLA-B27, for example, is present in up to 90 percent of people with ankylosing spondylitis. However, this antigen is also present in about 5 percent of people without any autoimmune disease, so the fact that this HLA is present cannot be used to confirm a diagnosis of the disease.

ASK THE EXPERT

ARTHRITIS

Joint inflammation and damage, which can cause pain and in some people disability.

What are the causes?

There are many types of arthritis:

- Osteoarthritis is the most common form of the disease, in which joint damage, often caused by stresses put on the joints, results in pain and stiffness (p. 145).
- Rheumatoid arthritis results when autoantibodies (antibodies produced by the immune system that are

directed against the body's own tissues) attack the synovium, the membrane lining the capsule that encases a joint, causing it to become inflamed and eventually resulting in damage to the bones of the joints (p. 150).

- **Septic arthritis** is caused by an infection (p. 152).
- **Reactive arthritis** is a reaction to an infection elsewhere in the body.
- **Juvenile chronic arthritis** (JCA) is a disease of childhood in which joint inflammation may be accompanied by symptoms in other parts of the body. The cause of JCA is unknown, but it is thought that genetic factors may be involved (p. 142).

The various types differ in the pattern of their symptoms, the joints most commonly affected, and the people they particularly tend to affect. Other types of arthritis include gout (p. 141), pseudogout (p. 149), and ankylosing spondylitis (p. 134). The skin disease psoriasis can be associated with arthritis, as can the inflammatory bowel condition Crohn's disease.

What are the symptoms?

The main symptom of arthritis is usually pain. This can be accompanied by swelling, stiffness, and restricted movement of the affected joint. In the case of rheumatoid arthritis, the pain and stiffness are worse in the morning. In some types, including rheumatoid arthritis, the affected joints feel warm to the touch. In reactive arthritis, the joints also appear reddened. Deformities may eventually develop, and walking and other activities can be affected.

In some types of arthritis, there are other symptoms in addition to joint problems. For example, individuals with septic arthritis tend to have a fever. A common feature of rheumatoid arthritis is anemia, which causes fatigue, and there is inflammation of other parts of the body, such as the heart and lungs. In reactive arthritis, there may be symptoms caused by infection elsewhere, for example, diarrhea if there is an infection of the digestive tract and pain when urinating if the urinary tract is the area affected.

How is it diagnosed?

It may be possible to identify the type of arthritis from a description of the symptoms and from which joints are affected. Blood tests and X rays may be performed as part of the assessment. Blood samples can be tested for signs of inflammation and for particular features that may be found in certain types of the disease. These include the antibody rheumatoid factor, which is often found in people with rheumatoid arthritis.

HELP YOUR DOCTOR HELP YOU

Diagnosing arthritis

Several features enable the doctor to diagnose arthritis and distinguish among the various types of the disease. If you have been having joint pains, it may be useful to think about the following questions before your appointment:

- *How long have you had the pains?*
- *Are your joints stiff, and, if so, are they worse at any particular time of the day?*
- *Which joints are affected?*
- *Do you have other symptoms that seem unrelated to your joints?*
- *Do you have a family history of joint problems?*

What are the treatment options?

Treatments often include painkillers and nonsteroidal antiinflammatory drugs. Steroid injections into an affected joint may be appropriate if the pain is severe. This is not appropriate for septic arthritis, in which the mainstay of treatment is high-dose antibiotics to clear the infection causing it. Surgery may be offered to drain the affected joint. Physical therapy is recommended when mobility is affected. In some types of the disease, replacement of a severely damaged joint may be the only way of relieving pain and improving mobility.

What is the outlook?

The prognosis varies by type of arthritis and among individuals. Symptoms can often be relieved by drugs and physical therapy, and joint replacements offer hope to many people with severe pain. Reactive arthritis tends to resolve completely over a period of a few months.

See also Living with arthritis, p. 113.

BONE TUMORS

Benign (noncancerous) or malignant growths in bones.

MALIGNANT TUMORS

Malignant tumors originating in bone (primary bone tumors) are rare and mainly affect children and young adults. Of these, osteogenic sarcoma is the most common; it affects approximately one person in a million each year. More common are secondary, metastatic bone tumors that spread from a primary cancer elsewhere—the lung, breast, prostate, or kidney in particular.

What are the symptoms?

Osteogenic sarcoma affects the ends of the long bones, especially in the leg above or below the knee. Swelling in the area may be accompanied by pain. In the case of metastases, there may also be symptoms caused by the original cancer. An affected individual may feel generally unwell and have a fever. Bone tumors can also cause pathological fractures—breaks resulting from a minor injury that would not cause a fracture in healthy bone.

How is it diagnosed?

Blood tests may include checking calcium levels, which are elevated in up to a fifth of people with bone cancer. Levels of alkaline phosphatase are also likely to be elevated. X rays may show areas of bone destruction, and bone scans (p. 108) show "hot" or bright areas if metastases are present.

What is the treatment?

Painkillers and nonsteroidal antiinflammatory drugs are taken to relieve pain caused by a tumor. Anticancer drugs accompanied by local radiation, and sometimes removal of the affected bone, may be appropriate for some patients with primary bone tumors. Some bone tumors depend on certain hormones for their growth and so antihormone drug therapy may be used. Metastatic tumors are often treated with local radiation, and treatment may be given for the original tumor.

What is the outlook?

In the case of primary malignant tumors, this depends on how soon the symptoms develop and the diagnosis is made, on the type of tumor, and whether treatment with anticancer drugs is appropriate. About a third of individuals with an osteogenic sarcoma will survive long term. Local radiotherapy for metastases aims to give a period of remission but cannot prevent tumors from recurring.

BENIGN TUMORS

Benign tumors are relatively common but in most cases are no cause for concern and require no treatment. However, some do cause problems, such as pain, slight bone deformities, or an increased liability to fracture.

Exostoses are outgrowths from bone. A common site is the heel, and removal may be required if the growth interferes with wearing shoes. Exostoses may also be present at the end of long bones, where they may prevent muscles from contracting properly and so may need to be removed. In rare cases, such growths become malignant.

Two benign tumors—osteoid osteomas and osteoblastomas—may be diagnosed on isotope bone scans. Surgical removal is an option if they are causing severe pain.

BUNION

A swelling next to the joint where the big toe meets the foot.

A bunion consists of a bursa, a fluid-filled sac, lying close to the joint between the big toe and the foot. Bunions are more common in women than in men.

What are the causes?

Bursae can be inflamed by friction over joints and bones. In the case of bunions, they are the result of wearing poorly fitting shoes that rub over the joint. A bunion is more likely to develop if there is an exostosis (a benign bony outgrowth) on the first metatarsal bone (the bone of the foot that connects with the big toe) or if the end of this bone is particularly prominent.

What are the symptoms?

A bunion may cause pain and in some cases deformity if the big toe is pushed out of alignment: It shifts toward the other toes. Difficulty in walking is common. The bursa may become inflamed, causing redness and additional pain. This is likely if the person continues to wear tight shoes that rub the area, causing an abrasion that allows infection to enter.

What are the treatment options?

For most people, well-fitting shoes and cushioning of the bunion by a soft pad is all that is required. For some individuals, surgery to remove the bursa and the bony prominence underlying it may be needed. Antibiotics may be prescribed if a bursa becomes infected.

What is the outlook?

Most people halt the progression of bunions by switching to wearing well-fitting, comfortable shoes. In severe cases, surgery is usually successful, but patients are advised to wear well-fitting shoes to prevent a recurrence. Bunions increase the risk of osteoarthritis (p. 145) later in life.

BURSITIS

Inflammation of a fluid-filled sac overlying a joint.

Bursas normally grow over a bone or joint to cushion against friction. The knee and shoulder are frequent sites.

What are the causes?

In most cases, the condition results from excessive friction over a bursa. Less commonly, a bursa may become inflamed as a result of infection or injury. In housemaid's knee, the prepatellar bursa, which allows the skin to move smoothly over the patella (the kneecap), becomes inflamed as a consequence of excessive kneeling.

Housemaid's knee can affect anyone who kneels excessively at work, such as tilers, landscape gardeners, and carpet fitters.

What are the symptoms?

The bursa becomes painful and swollen; movement of the affected joint may be limited. In housemaid's knee, a swelling will be noticed in front of the patella. A diagnosis is made by physical examination.

What are the treatment options?

The most important part of treatment is to rest the affected joint. For example, if the knee is affected, kneeling should be avoided until the inflammation has subsided, and then kneeling pads can be worn for protection and time spent kneeling should be reduced, if possible. Nonsteroidal antiinflammatory drugs may help to relieve inflammation.

If the symptoms persist, the doctor may drain the fluid from the bursa and inject it with a corticosteroid drug. In cases where the condition persists or frequently recurs, removal of the bursa may be recommended. Bursitis caused by infection is treated with antibiotics and often surgery (irrigation and debridement).

CERVICAL SPONDYLOSIS

A common condition in which wear and tear of the vertebrae and discs of the cervical spine in the neck results in stiffness and pain.

The condition tends to occur in people over age 40; men are affected more than women. Young people may develop the condition following a neck injury.

What are the symptoms?

Affected individuals often feel a pain in the neck, which spreads across the shoulders and down the arms and may restrict movement. Deformities of the cervical spine may develop and cause nerve compression, resulting in tingling and sometimes numbness in the hands and arms. With sudden neck movements, these deformities may also compress blood vessels that supply the brain, causing transient dizziness.

How is it diagnosed?

Cervical spondylosis is diagnosed by X ray. In some cases, where there are no symptoms, the disorder may be picked up on X rays taken for another reason, perhaps following an injury. An MRI (p. 107) may be performed to look at the vertebrae and discs in more detail.

What are the treatment options?

Painkillers or nonsteroidal antiinflammatory drugs may be taken. Rest and a warm compress may also help relieve the pain. When the symptoms have subsided, exercises are recommended to improve neck mobility. Rarely, if the pain is very severe, an operation to fuse the affected vertebrae in position may be considered.

What is the outlook?

The changes in the spinal column may remain stable for many years or may progress slowly. In most cases, the symptoms of cervical spondylosis are mild and resolve with rest, heat, and analgesia. Surgery is only suggested in the more serious cases if mobility is severely limited; fusion of the vertebrae aims to provide lasting relief.

CHONDROMALACIA PATELLAE

Softening of the cartilage on the back of the patella (kneecap) resulting in pain in the front of the knee.

This condition particularly affects teenagers. Excessive strain placed on the joint between the lower end of the femur and the patella results in softening, swelling, and deformity of the cartilage between the two bones.

What are the causes?

Undue strain may be placed on the joint as a result of frequent episodes of strenuous exercise, such as running long distances. Chondromalacia patellae may also be the result of repeated injuries to the knee.

What are the symptoms?

Pain in the knee is the main problem, together with stiffness after sitting for a long time. In more severe cases, there may be crepitus (a crackling sound caused by the bone rubbing against the damaged cartilage) and a clicking sensation when the knee is bent and straightened.

How is it diagnosed?

Following a thorough examination of the knee, X rays will be taken. An MRI may also be arranged when the knee joint needs to be examined in more detail.

What are the treatment options?

Painkillers and nonsteroidal antiinflammatory drugs will help to relieve the pain. If the condition is not severe and the X rays look normal, the specialist will keep the patient under observation with regular follow-up exams, perhaps every six months. Gentle exercise targeting the muscles of the thigh and around the knee may be recommended to reduce the strain placed on the joint.

If the condition is severe, arthroscopy (p. 119) may be recommended, during which the joint interior will be examined and any irregularities on the surface of the cartilage removed.

What is the outlook?

In mild cases, the condition is likely to resolve. However, if the condition is severe, cracks in the cartilage may begin to appear and the symptoms will persist. In these cases, people are at high risk of developing osteoarthritis (p. 145) of the patellofemoral joint—the joint between the thigh bone and kneecap—in later life.

DISLOCATION OF JOINTS

A common disorder in which the joint surfaces slip across each other so that they are no longer in contact.

What are the causes?

Joint dislocation is the result of a force applied to a joint. Falls and injuries when playing contact sports are common causes. Shoulder and finger joints are particularly vulnerable to dislocation. In some cases, the joint surfaces remain partly in contact; this partial dislocation is known as subluxation. Sometimes a dislocation is accompanied by a fracture close to the joint, a so-called fracture dislocation. The ligaments that hold the bones in place will also be damaged, as may the capsule that surrounds the joint.

What are the symptoms?

Dislocation causes severe pain and deformity of the joint. This is usually accompanied by swelling and restriction of movement. There may be local bruising of the skin.

How is it diagnosed?

The condition is usually clear from the symptoms and the appearance. However, X rays are always taken to confirm a diagnosis and assess other injuries, in particular fractures.

What are the treatment options?

Painkillers may be required for the first few days following the injury. Subluxation usually requires no treatment other than pain relief, and the joint normally becomes fully stable when the tissues around it have healed. Full dislocation requires the joint to be manipulated back into its proper position (reduction) and then to be kept in position (immobilized) for up to six weeks to allow the supporting structures to heal; rarely, an operation may be needed to repair the damage. After immobilization, physical therapy may be needed to restore mobility of the joint. If there is a fracture, the detached bone may need to be fixed to the main part of the bone and immobilized while it heals.

What is the outlook?

In most cases, dislocation does not cause any long-term problems. Sometimes, however, the joint becomes more liable to dislocate, and the problem may recur. Fractures associated with dislocations usually heal well.

DUPUYTREN'S CONTRACTURE

Thickening and scarring of the tissues in the palm of the hand causing the fingers, particularly the ring and little finger, to become permanently bent.

This fairly common condition affects men more than women and may run in families.

What are the causes?

Dupuytren's contracture is caused by thickening and shortening of the fibrous tissue in the palm of the hand, the palmar fascia. The fingers are drawn into the palm, producing a deformity. This is sometimes associated with long-term use of vibrating tools, such as drills. For unknown reasons, it is associated with certain diseases, particularly diabetes mellitus, and with medictions used to treat seizures. Long-term alcohol abuse is also a risk factor.

What are the symptoms?
Dupuytren's is a painless condition; the main problem it causes is restriction of certain activities, because of shaking hands. Both hands are often affected. Rarely, Dupuytren's of the soles of the feet develops.

What are the treatment options?
In mild cases, no treatment is necessary. If the deformity causes problems, surgery may be recommended to remove the thickened tissues and straighten the affected fingers as much as possible. The deformity can, however, return following surgery.

FIBROMYALGIA
A condition in which there are widespread muscle pains, often accompanied by stiffness.

Fibromyalgia is more common in women. The cause is unknown, but it often develops during times of stress.

What are the symptoms?
The muscle aches are persistent, and the pain is often worse at particular points around the body, especially around the neck and shoulders, elbows and knees. Fatigue is common, as are headaches and problems sleeping. An affected individual may also be depressed and anxious.

How is it diagnosed?
There are no specific tests to diagnose fibromyalgia, although investigations may be arranged to rule out other causes of the pain.

What are the treatment options?
Painkillers and nonsteroidal antiinflammatory drugs may be taken intermittently to relieve the pain. Low doses of certain types of antidepressants may be prescribed for their pain-relieving properties. Exercise is an important part of any treatment program. Massage and physical therapy may also be helpful. Injections of anesthetic drugs or corticosteroids into the points where the pain is most severe are sometimes helpful. In some cases, acupuncture may help to relieve the pain.

What is the outlook?
The condition often completely resolves, or at least the symptoms can be controlled sufficiently to maintain an active life. In a few cases, fibromyalgia restricts activities.

FRACTURES
Breaks in bones, often as a result of trauma.

Fractures are common. In addition to damage to bones, there may be damage to the blood vessels, muscles, and soft tissues around the fractured bone. They may be classified as open or closed, depending on whether there is an open soft tissue wound or whether the skin has remained intact. Fractures are often given names that describe the fracture line or the fragments of bone at the site of the break:

- **Transverse fractures** show a line of breakage across the bone, usually as a result of a direct blow.
- **Spiral fractures** are the result of a twisting injury, which produces a spiral fracture line in the bone.
- **Comminuted fractures** have more than two fragments.
- **Crush fractures** leave no obvious pieces to join back together, since the bone is crushed rather than broken; crush fractures particularly affect the vertebrae of the spinal column.
- **Greenstick fractures**, where the bone is not completely broken and its outer layers remain intact, usually occur in children.

What are the causes?
A fracture may result from a direct blow to a bone or from a twisting injury in which the bone twists on its axis. Repeated stress placed on a bone can lead to a fracture; the tibia (shinbone) is a common site for stress fractures in long-distance runners. Bones may also be more prone to fracture if they are weakened, for example by a malignant tumor or by osteoporosis (p. 146), a common cause of crush fractures of the vertebrae.

What are the symptoms?
The symptoms depend in part on the site and severity of the fracture, as well as on the extent of the damage to the soft tissues around the break. The following are often present following a fracture:

- pain and tenderness at the site of the fracture;
- swelling in the affected area;
- bruising, which may take several hours to appear;
- crepitus, a grating sensation as the broken ends of bone rub together; and
- deformity of the affected limb, often accompanied by abnormal movement.

The symptoms also depend on the position of the bone fragments; if they have not been displaced out of position, for example, there may be no deformity.

Bone fractures can lead to various complications. Some develop soon after the fracture. These early complications include severe hemorrhage and infections, as well as damage to nearby body structures. For example, a fractured rib may occasionally lead to lung damage.

In the long term, fractures may result in bone deformity if the bones heal in an incorrect position (mal-union) or if they don't heal at all (nonunion). If mal-union follows a fracture, extra stresses may be put on nearby joints, causing osteoarthritis in later years. Another possible long-term complication is aseptic necrosis, a serious condition in which a fracture severs the blood supply to an area of bone, causing the bone tissue to die off gradually. This damage may take many years to develop. The head of the femur (the thigh bone) is particularly vulnerable, and pain and stiffness of the hip joint are the eventual result.

What is the treatment?

Pain relief is important, particularly in the early stages, but the main aim of treatment is to help the bone fragments reunite strongly in their original position. If there is an open wound, the area is cleaned first. Then the fragments of bone are returned as close as possible to their original position (reduction of the fracture) and fixed in place (immobilization), so that they can heal with a minimum of deformity. In some fractures, the broken ends are not displaced and simply need to be held in position. Various methods may be used to immobilize a fracture, including traction, plaster casts or fixing with metal screws and plates. Some fractures, for example of a single rib, usually require only pain relief. Finally, physical therapy is often needed to mobilize the affected area and strengthen the muscles.

What is the prognosis?

In general, fractured leg bones take about 16 weeks to heal; other bones take around 8 weeks. Children's bones tend to heal more quickly. Healing may be delayed by infection or problems in keeping the bone fragments in position. An impaired blood supply may also prevent the bone ends from healing (nonunion).

FROZEN SHOULDER

A troublesome disorder in which pain is followed by stiffness and restricted movement of the shoulder.

Frozen shoulder is common, particularly in women over the age of 40.

What are the causes?

The disorder may result from an injury to the shoulder. A period of immobility of the arm—following a stroke, for example—may also be the cause. In most cases, no specific cause can be identified.

What are the symptoms?

Frozen shoulder tends to occur in stages. For the first six months or so, movement of the shoulder joint is restricted by pain, which is often severe. As the pain begins to settle, stiffness becomes the main feature, still accompanied by restricted movement; this stage can last for up to a year. The last six months are the period of recovery, during which time the stiffness resolves and movement gradually returns.

How is it diagnosed?

Frozen shoulder is usually diagnosed from an assessment of the symptoms and a thorough examination of the shoulder. Other tests, such as X rays, may sometimes be arranged if there is any question of another disorder.

What are the treatment options?

Nonsteroidal antiinflammatory drugs are recommended for the painful phase of the condition. Later, physical therapy and hydrotherapy may help reduce the stiffness and improve mobility.

What is the outlook?

Recovery is slow, and even at the end of the recovery period there may be persistent restriction of movement.

GANGLION

A ganglion consists of a collection of thick fluid that produces a harmless swelling in the sheath surrounding a tendon or in the tissues around a joint. Common sites are the inside of the wrist and the back of the hand.

What are the causes?

Synovial fluid, which is found within joints and acts as a lubricant, is released into tissues outside a joint. The reasons for this are unclear.

Ganglia are three times more common in women than men and usually occur in early adulthood.

What are the symptoms?

Ganglia vary in size; small ones may cause no symptoms, but a large ganglion can interfere with movement. They are usually painless, although those on the hand and wrist may ache after use of the hand and may eventually become painful. Some ganglia cause compression of nerves, although this is relatively rare.

What are the treatment options?
Ganglia may disappear without treatment, sometimes as the result of an accidental blow. Small swellings can often be left, but if a ganglion interferes with movement or causes nerve compression, it should be removed surgically.

What is the outlook?
Ganglia often recur, even following excision. Following an operation, there is likely to be swelling in the area, which may take up to six months to resolve.

GOUT
A common condition in which uric acid crystals are deposited in a joint, most commonly in the big toe, causing arthritis and severe pain.

Uric acid is a product of the breakdown of cells and proteins in the body and is found in high levels in the blood of people with gout. Gout affects men more than women and is rare before puberty and in women before menopause. It often runs in families. A similar condition, pseudogout (p. 149), affects women more than men.

What are the causes?
There are various causes of high uric acid levels: The waste product may be produced excessively or the kidneys may fail to dispose of it adequately. Certain conditions, including obesity, a high protein diet, and diabetes mellitus, are associated with high uric acid levels. But not everyone with raised uric acid levels will develop gout. Certain factors can predispose to the condition: dehydration, which may result from alcohol abuse; use of diuretic medication, for example for heart problems; and cell breakdown as a result of certain types of cancer treatment. Often there is no obvious reason why gout develops.

What are the symptoms?
The pain of gout can be severe and is accompanied by redness and swelling around the affected joint. In most cases, this is the big toe. In long-term gout, crystal deposits may cause small lumps called tophi in the earlobes and the tissues of the hands and feet.

> ### EXPERIENCING GOUT
> ***I woke one morning with a terrible throbbing pain in my big toe. When I looked at it, it was swollen and red.***
>
> *I wondered if this could be gout—I'm in my 50s now and have always enjoyed a few beers. I went down to the doctor. She took a blood sample from me but said that she was pretty sure that it was gout. I'm pleased to say my symptoms began to settle within a few days of taking nonsteroidal antiinflammatory drugs, although the toe was very painful for a time, and difficult to walk on. I was also advised to eat cherries; these should lower uric acid levels in my body. I've now made some changes to my lifestyle—I've cut down my visits to the bars and stick to a glass of wine with my evening meal and have lost a few pounds—to try to make sure it doesn't come back.*

How is it diagnosed?
The diagnosis is usually clear from the symptoms and the examination. Blood uric acid levels are measured, and occasionally a sample of fluid may be drawn from the joint and examined under a microscope for uric acid crystals.

What are the treatment options?
Nonsteroidal antiinflammatory drugs will relieve the symptoms in many cases. High doses are given for about two days and then the dose is reduced for perhaps another week. Other drugs, such as colchicine, are used for individuals unable to tolerate nonsteroidal antiinflammatory drugs. If the condition is recurrent and the symptoms are severe, drugs that lower uric acid levels may be considered. They should not be taken until at least four weeks after the last attack. Lifestyle changes may be recommended, including reducing alcohol intake, reducing the intake of calories and cholesterol, and avoiding foods rich in purines such as some types of fish, shellfish, and spinach. The end product of purine metabolism is uric acid.

What is the outlook?
Gout often recurs, but the likelihood can be reduced by dietary and lifestyle changes. Recurrent gout may eventually cause permanent joint damage.

JUVENILE CHRONIC ARTHRITIS

A disease of childhood in which joint inflammation is sometimes accompanied by symptoms in other parts of the body.

The precise cause of juvenile chronic arthritis (JCA) is unknown, although it can run in families, suggesting that genetic factors are involved. JCA usually lasts at least six weeks and can take one of three forms:

- systemic, also known as Still's disease;
- polyarticular, in which many joints are involved; and
- pauciarticular, in which a few joints are involved.

What are the symptoms?

Still's disease is the rarest form of juvenile chronic arthritis. It affects boys and girls equally at any age. A child with the disease may have a variety of symptoms, including fever, weight loss, and abdominal pain, but, at least initially, there may be no joint symptoms. Later, pain, swelling and stiffness of joints are likely to develop.

In polyarticular JCA, the main symptoms are pain, swelling, and stiffness of large and small joints, which is particularly severe in the morning. Girls are affected more than boys.

Pauciarticular JCA involves fewer than five joints, particularly the knees, ankles, and elbows; often, however, only one joint is affected. It is the most common form of JCA, and girls under the age of four are most often affected. Children with this form of JCA are at risk of developing chronic iridocyclitis, inflammation of structures in the eye which, if untreated, can eventually lead to blindness.

How is it diagnosed?

Blood tests will check whether the erythrocyte sedimentation rate is elevated (ESR, p. 109). This indicates that inflammation is present but does not confirm an eventual diagnosis. Affected children are anemic, and X rays will show signs of inflammation and damage in the joints.

What are the treatment options?

Treatment aims to reduce inflammation, relieve pain, and keep the joints as mobile as possible. To this end, nonsteroidal antiinflammatory drugs are combined with physical therapy, incorporating exercises and hydrotherapy. In severe cases, corticosteroid and other antiinflammatory drugs may be taken orally or injected into affected joints. Treatment for iridocyclitis includes corticosteroid eye drops or oral steroids in severe cases. The affected child and the family will be offered support and advice in how to cope with the disease.

What is the outlook?

In most cases, the prognosis is good and the arthritis clears. However, there is a risk of long-term arthritis with the systemic form. The risk of developing chronic iridocyclitis is a concern in pauciarticular JCA, and regular checkups by an ophthalmologist are essential. Treatment for this serious eye disorder is usually successful.

KYPHOSIS

Excessive curvature of the upper part of the spine resulting in a stooped posture.

A mild form of this curvature is present in the normal spine (p. 22), but in kyphosis the curvature is pronounced enough to affect posture.

What are the causes?

Mild kyphosis may simply be the result of bad posture: Some people develop a stoop at an early age, perhaps because they are conscious of being tall. Other causes of kyphosis include osteoporosis (p. 146), which can result in vertebral collapse and kyphosis, and ankylosing spondylitis (p. 134).

Slouching and poor posture can stretch spinal ligaments, increasing the natural curve of the spine. This postural kyphosis usually begins in adolescence and often resolves as posture improves.

What are the symptoms?

Kyphosis may be accompanied by back pain, resulting from the underlying cause. Pain and discomfort may also be caused by strain put upon the back muscles.

How is it diagnosed?

Kyphosis can be seen in a physical examination. The severity and possibly the cause can be judged from lateral X rays (X rays taken from the side).

What are the treatment options?

Treatment and prognosis depend on the underlying cause. Pain relief and physical therapy to strengthen the back muscles may be recommended. In severe cases, an operation may be performed to fuse the affected vertebrae. Treatment to counter osteoporosis may delay the progression of the problem.

LIGAMENT INJURIES

Damage to the tough bands of tissue that keep joints stable by holding the bones in position.

Ligament injuries are common; the knees and ankles are especially vulnerable. In a sprain, there is a ligament tear, but the affected joint remains stable. In complete ruptures, the ligament is severed and the joint is unstable.

What are the causes?

Sports accidents are the most common cause. Soccer players often damage their cruciate ligaments, the fibrous bands that prevent the knee from bending or straightening excessively. Ligament injuries are more likely during exercise if there has been no warm-up. These injuries also often occur as a result of falls in which ligaments are overstretched.

What are the symptoms?

The symptoms come on suddenly and may include

- pain that is worse when the affected joint is moved,
- swelling and bruising in the affected area, and
- restricted movements of the joint.

How is it diagnosed?

The diagnosis is usually clear from a description of the symptoms and an examination. However, if there is any doubt, X rays may be taken to exclude a fracture.

What are the treatment options?

Painkillers are likely to be needed. It is important to avoid putting any strain on the affected joint until the injury has healed. For more severe injuries, a plaster cast may be necessary for about six weeks. For complete rupture, surgical repair is sometimes recommended. In all cases, physical therapy can help to restore mobility and strength.

What is the prognosis?

Sprains usually heal well, although there is a risk that they will recur. Complete ruptures often need aggressive treatment such as surgery, but the outcome is generally good.

LOWER BACK PAIN

Discomfort in the part of the back below the waist.

Pain in the lower back is an extremely common problem, most often caused by a muscle or ligament strain, but sometimes the result of a more serious underlying problem.

What are the causes?

It is easy to strain the muscles of the back. Poor posture and a sedentary occupation predispose people to back problems. Lifting a heavy object with the wrong technique is a common cause. Even seemingly trivial sudden movements can cause a muscle or ligament strain around the lumbar spine. Doing certain types of activities without warming up the muscles of the back first, or lifting after sitting for a long period, can also cause back strains. Lower back pain is common during pregnancy because the weight of the baby and the postural changes that are necessary to accommodate it. Back pain can also be caused by a prolapsed disc (p. 148). This can develop gradually or suddenly as a result of a sudden movement.

Long-term disorders can be associated with lower back pain. Osteoarthritis (p. 145) of the spine is a common cause. Less common causes include ankylosing spondylitis (p. 134) and bone tumors that have spread from a cancer elsewhere in the body (p. 135). Sometimes lower back pain is the result of a problem within the body, such as infection in the urinary system.

What are the symptoms?

Lower back pain may develop suddenly or it may come on gradually. It may be localized or spread across the whole lower back and may vary from mild to severe in intensity. Stiffness and restriction of movement are common.
The pain may spread down the leg. If there is pressure on nerves branching off the spine or on the spine itself, there may be also be tingling and sometimes numbness in the legs. In severe cases, there may be weakness in the legs and problems with bladder and bowel control.
This situation requires urgent medical attention: Nerve pressure may lead to irreparable damage.

How is it diagnosed?

A full physical examination may be all that is required. However, X rays may be taken to exclude certain causes of back pain. CT scanning or an MRI (p. 107) may be used, in particular to diagnose a prolapsed disc.

What are the treatment options?

Resting on a firm bed or lying on the floor may be helpful for muscle and ligament strains, but such immobility should not last for more than 48 hours. Painkillers and nonsteroidal antiinflammatory drugs may provide relief, as may applying heat to the area in the form of a heating pad or hot water bottle wrapped in a towel.

If lower back pain is persistent, medical advice should be sought. In most cases, conservative treatment—including pain relief, nonsteroidal antiinflammatory drugs, and physical therapy—will be recommended rather than surgery.

What is the outlook?
Lower back pain is often short-lived, but it tends to recur. Appropriate lifting techniques and good posture will reduce the risk, as will regular exercises to strengthen the back muscles. For a few people, lower back pain is a persistent and debilitating problem. It is important for them to remain as active as possible and to seek expert advice, perhaps from a physical therapist, on how to achieve this.

MENISCAL TEARS
Damage to the menisci, the discs of cartilage within the knee that act as shock absorbers and help the knees bear the heavy load of the body.

What are the causes?
Meniscal tears are common in athletes. They may result from a seemingly trivial injury, such as twisting the knee when it is bent and the foot is on the ground. They may even occur when turning over in bed. In some cases, there is no obvious injury, and an individual may not be aware that damage has occurred until the symptoms develop.

What are the symptoms?
With a sudden injury, there may be a sharp pain, followed by swelling of the knee. Fragments of the meniscus may loosen and move into the joint. Transient locking, a condition in which the knee is temporarily "locked" into a straight position because the meniscal fragments prevent the bones from moving smoothly over each other, is common. It may be impossible to place the full body weight on the knee.

How is it diagnosed?
The doctor will often suspect the diagnosis from a description of the symptoms and examination. It must be confirmed either with an MRI or by arthroscopy (p. 119).

What are the treatment options?
Any loose fragment must be removed while preserving as much of the affected meniscus as possible. This may be carried out arthroscopically or less commonly by open surgery, in which the knee joint is opened up.

Physical therapy may be recommended to improve the strength of the muscles around the knee and the mobility of the joint. A patient can return to heavy activity two weeks after arthroscopic treatment, but it may be three months before heavy work can be resumed following open surgery.

What is the outlook?
Surgery to repair torn menisci usually produces good results, but in some cases there may be an increased risk of developing osteoarthritis (p. 145) of the knee later on.

MUSCLE STRAINS
Injuries to muscles caused by overstretching their fibers.

Muscle strains are common and rarely cause long-term problems. Common sites are the back, the limbs, and the muscles of the abdominal wall. Sometimes some of the fibers of the injured muscle are actually torn.

What are the causes?
Various actions, including something as simple as reaching for an object on a high shelf can result in muscles strains. Muscles are often strained when playing sports. Back muscles can be "pulled" by standing up suddenly, and both back and abdominal muscles may be strained by poor lifting technique. A warm-up session at the beginning of a period of exercise reduces the risk of strains.

What are the symptoms?
The main problem is pain, often accompanied by stiffness and limited movement of the affected area. Bruising is also common but may take several hours to appear.

What are the treatment options?
Muscles strains should be treated by rest, and an ice pack should be applied for about five minutes to reduce the swelling and pain. The area should be gently but firmly compressed and, if possible, elevated (p. 96). Nonsteroidal antiinflammatory drugs should help relieve the symptoms. If symptoms persist for more than about two days, medical advice should be sought. The doctor will take an X ray if there is any possibility of a fracture. Physical therapy may be recommended.

What is the outlook?
Muscle strains usually heal well. However, a muscle tear may require surgery to rejoin the severed fibers.

OSTEOARTHRITIS

Worsening damage to the cartilage that lines joints, which may cause pain and stiffness.

This is the most common type of arthritis, and it develops as people age; most people in their 60s will show some evidence of the condition on X rays, even if they have no obvious symptoms. Women tend to be affected more than men. In osteoarthritis, it is the cartilage lining the bones of the joints that is damaged first. Then the bone becomes thickened and osteophytes (bony growths) develop. Any joints may be affected, but the knees and hips, which bear the weight of the body, are particularly vulnerable.

Osteoarthritis is the most common form of joint disease, affecting more than 15 million Americans and 5 million people in the UK.

What are the causes?

The disease often develops without any obvious cause. However, anything that puts excessive strain on particular joints over a long period can cause osteoarthritis. Ballet dancers, for example, commonly develop osteoarthritis of the joints in their toes and ankles fairly early in life. Osteoarthritis is also more likely to develop in people who are overweight. In such cases, the knees and hips (the weight-bearing joints) are especially at risk.

What are the symptoms?

Any joints can be affected. Pain and stiffness are the main symptoms, often accompanied by swelling. Crepitus (a crackling sound produced when bone rubs over the damaged cartilage) may also be noticed. As the disease worsens, affected joints can become unstable. This is particularly noticeable when the hip or knee is affected and walking difficulties develop.

Osteoarthritis of the joints in the hands can also be debilitating. The joints of the hands become enlarged, and the hands may appear misshapen. Muscles around joints affected by osteoarthritis eventually weaken and become wasted (thinner).

How is it diagnosed?

There are no specific blood tests to diagnose the condition. X rays may appear normal until the joint damage is quite severe, but they may be useful to exclude other possible causes of the pain.

What are the treatment options?

Weight loss will be recommended for those who are overweight to reduce the stress placed on the affected joints. Exercises to strengthen muscles around the joints may be helpful. Swimming is an excellent way to do this, exercising the muscles while the body's weight is supported. Nonsteroidal antiinflammatory drugs applied in gel form may give some relief, as may heat pads held against aching joints. For more severe symptoms, painkillers and courses of nonsteroidal antiinflammatory drugs may be used. Corticosteroid injections into joints may be given, but frequent injections into one joint are avoided. In severe cases, joint repair or replacement may be offered as the only way to reduce the pain and restore mobility. In very severe cases, certain joints, such as those within the wrist, can be fused rigid. This may relieve the pain, but it leaves the patient with no movement in the affected joint.

What is the prognosis?

Osteoarthritis cannot be cured, but various measures can help relieve pain and improve the function of the affected joints (p. 113). Joint replacement has a good prognosis, giving relief from pain and improved mobility for many years following surgery.

OSTEOMALACIA

Softening of the bones as a result of vitamin D deficiency; rickets is the childhood form.

Vitamin D is necessary for the body to be able to obtain calcium (essential for strong, healthy bones) from food. Vitamin D is found in eggs, green vegetables, fish, and milk.

What are the causes?

Osteomalacia may result from a deficiency of vitamin D in the diet or from a lack of exposure to sunlight (vitamin D is produced in the body when it is exposed to sunlight). The latter may occur in elderly people who are housebound. Impaired absorption of vitamin D in the gut also leads to a deficiency, which may eventually cause osteomalacia or rickets. Other causes of osteomalacia include kidney failure and inherited disorders in which vitamin D processing is impaired.

What are the symptoms?

The main symptoms of osteomalacia are aching and tenderness of the bones. In childhood rickets, bone deformities are present, in particular the characteristic bowed legs. Growth is impaired. In both osteomalacia and rickets, bones are liable to fracture easily.

How is it diagnosed?

Blood tests, including those for calcium levels, are likely to be performed. X rays show softened bone that is lacking in calcium. Occasionally, a tiny sample will be taken from the hip bone to confirm the diagnosis.

What are the treatment options?

Treatment aims to treat the cause wherever possible, for example, by increasing the amount of vitamin D in the diet and the exposure to sunlight where appropriate. Vitamin D injections may be needed, in which case calcium levels need to be monitored.

What is the outlook?

With appropriate treatment, the prognoses for osteomalacia and rickets are good. However, the deformities of severe rickets may remain.

OSTEOMYELITIS

Inflammation of bone caused by an infection.

This may be an acute disease, which develops suddenly, or a chronic, persistent condition, which comes on more gradually. Acute osteomyelitis particularly affects children.

What are the causes?

An infection may reach a bone from the blood if it spreads from an infection in another part of the body. Alternatively, infections can invade bones directly following a fracture or occasionally after bone surgery.

What are the symptoms?

In the acute form, the affected individual feels unwell with a fever and severe bone pain. Children are likely to be reluctant to move the affected area. Left untreated, the infection spreads within the bone, and pus may eventually leak from the overlying skin. Alternatively, pus may leak into a nearby joint, causing septic arthritis (p. 152). Persistent infection is called chronic osteomyelitis. Symptoms of this include pain and weight loss. Often there is a channel leading from the affected bone to the skin surface (known as a sinus), which discharges pus. This may last for many years. Chronic osteomyelitis may also be caused by tuberculosis (TB).

How is it diagnosed?

Blood is taken to look for signs of infection and then to identify the infection responsible. X rays and an MRI may be performed to locate the infection site.

What are the treatment options?

Acute osteomyelitis is treated with rest and a combination of intravenous antibiotics. If the infection persists, a hole may be drilled in the bone to release the pus. Oral antibiotics will probably be necessary for several months to prevent further infection.

Chronic osteomyelitis can be difficult to treat. Extensive surgery may be required to remove the dead bone tissue, and this will be combined with a long course of antibiotics. Once the infection has been cleared, additional surgery may be required to graft the affected area of bone, replacing it with bone from elsewhere in the body.

What is the outlook?

With appropriate treatment, the outlook for acute osteomyelitis is good. Chronic osteomyelitis is more difficult to treat, and long-term therapy is needed to eradicate the infection.

OSTEOPOROSIS

Loss of bone tissue, resulting in weakened bones that are liable to break easily.

Throughout life, bone is continually being broken down and replaced. As we age, however, the rate at which bone is made becomes slower than the rate at which it is broken down. Thus a loss of bone tissue (osteoporosis), albeit gradual, is a normal part of the aging process.

Some factors trigger the development of osteoporosis at an earlier age than normal or may speed up the process. In both cases, severe osteoporosis in later life can result. This causes many fractures in the elderly, particularly of the vertebrae, wrists, and hips.

What are the causes?

Women are particularly at risk of developing osteoporosis after menopause. Prior to that they are protected by the high levels of the hormone estrogen, which is needed for new bone to be laid down. Early menopause therefore increases the risk of severe osteoporosis developing later in life. Men gradually develop osteoporosis later in life because of the decreasing levels of androgens (male sex hormones).

Can osteoporosis be prevented?

There is no way to eliminate all the risk factors, but adopting a healthy lifestyle at an early age will go a long way toward reducing the risks. It is what you do early in life that will make the biggest impact, but doing the following at any time in life can only be beneficial:

- Eat a healthy, balanced diet with adequate amounts of calcium and vitamin D.
- Do not smoke.
- Get regular exercise.
- Keep your alcohol intake within the recommended limits (21 units per week for men and 14 units for women, with at least one or two alcohol-free days every week).

Women may also wish to consider taking hormone replacement therapy after menopause.
Bone dual X-ray absorptiometry may be recommended for those known to be at increased risk of osteoporosis (individuals with a family history, for example).

ASK THE EXPERT

Certain factors increase the risk of developing osteoporosis. These include being persistently underweight, smoking, lack of exercise, excessive alcohol, and a diet very low in calcium. A family history also increases the risk. For unknown reasons, white and Asian women are at increased risk of developing the disease. Certain drugs are also associated with osteoporosis. These include corticosteroids and certain anticancer drugs. Other associated diseases include type I diabetes mellitus (the type treated by insulin), chronic liver disease, and rheumatoid arthritis (p. 150).

What are the symptoms?

Osteoporosis itself does not cause symptoms; these are caused by the complications of the disease. Pain is caused by fractures of the weakened bone. Severe pain in the back may occur as a result of vertebral fractures, which may be accompanied by loss of height and kyphosis (p. 142).

How is it diagnosed?

Blood tests may be used to look for an underlying cause. X rays may pick up fractures caused by osteoporosis, but they cannot be used to assess bone density. This task is performed by bone dual X-ray absorptiometry (DXA; p. 108).

What are the treatment options?

An appropriate diet with adequate calcium and vitamin D is needed. Exercise is recommended to maintain bone density. Giving up smoking is also important, because smoking may speed up the process of osteoporosis.

Various drug treatments are available. Hormone replacement therapy has been shown to reduce the risk of fractures in women after menopause. The decision to take HRT should be an informed one based on knowledge of its benefits, and also of the possible side effects. Drugs called bisphosphonates (p. 111) slow the breakdown of bone; some have been shown to increase bone mass and reduce the risk of fractures.

What is the outlook?

In the elderly, fractures may be life-threatening if they are followed by complications such as pneumonia. The risk of developing severe osteoporosis or of developing the condition earlier in life than normal can be reduced by lifestyle measures (p. 89) and by drugs (p. 110) where appropriate.

PAGET'S DISEASE OF BONE

A disease of middle to old age in which normal bone turnover is disrupted, with areas of the skeleton becoming thickened yet weak.

Bone renewal is an ongoing process; old bone is continually being broken down to be replaced by new. In Paget's disease, this process is excessive in areas of the skeleton, which become weakened and liable to fracture.

What are the causes?

The cause of Paget's disease is not yet known. It is more likely to affect people in some parts of the world than others (it is rare in Japan, common in Australia and the United States, and most common in the UK). It is thought that the cause may be viral, perhaps measles or the canine distemper virus. Evidence suggests that such a virus may be contracted early in life and then cause Paget's disease many years later.

What are the symptoms?

In many cases, Paget's disease produces no symptoms and is picked up as an incidental finding on X rays taken for another reason. Otherwise, the main symptom is pain in the affected area of bone or in a nearby joint, caused by osteoarthritis that has developed as a consequence of the disease. Deformities can eventually develop, often in the weight-bearing bones of the legs, but also in areas such as the skull and the spinal column, which can become curved.

Hearing problems may result from pressure of thickened areas of the skull on nerves to the ear. Pressure may be exerted on the spinal cord by deformed vertebrae. Other nerves around the body may also be compressed by deformed bone.

Blood supply to affected areas of bone increases, and in the elderly, this may put a strain on the heart, causing eventual heart failure, of which the main symptom is breathlessness.

How is it diagnosed?

Following an examination, X rays may be taken to confirm the diagnosis. Blood levels of a substance called alkaline phosphatase will also be checked This substance indicates that bone turnover is taking place, so in Paget's disease the levels are high. This test helps confirm the diagnosis but can also be used to monitor the severity of the disease and any improvements in response to treatment. These tests may be followed by a bone scan (p. 108) in order to assess the extent and severity of the disease.

What are the treatment options?

Painkillers or nonsteroidal antiinflammatory drugs are often helpful. However, if the pain persists or there are other problems, such as nerve compression or deformity, treatments that target the disease itself by reducing the excessive bone turnover may be considered. Such treatments may also be appropriate for younger people to delay progression of the disease and the development of deformities. Surgery may be performed to replace joints damaged by osteoarthritis or to treat fractures.

What is the outlook?

Paget's is a progressive disease. Drug treatments may be able to slow this progression, although existing deformities will not be reversed.

PROLAPSED DISC

A common condition in which one of the intervertebral discs of the spinal column bulges backward between two vertebrae.

The intervertebral discs are made up of a tough outer layer (the annulus fibrosus) encasing a softer center, known as the nucleus pulposus. In a prolapsed disc, that soft center bulges backward, distorting the disc, which may put pressure on nerves as they emanate from the spine (called nerve roots) or on the spine itself.

The lumbar spine is most commonly affected, in particular the disc between the fourth and fifth lumbar vertebrae (p. 22) or between the fifth lumbar and first sacral vertebrae in the lower part of the back. Higher parts of the vertebral column are sometimes affected, the cervical spine in particular.

Sometimes the annulus fibrosus ruptures, leaking disc contents into the spinal canal. This may cause worsening of the symptoms, but, strangely, the symptoms sometimes improve following this so-called disc rupture or herniation.

What are the causes?

Disc prolapse is often the result of a minor injury, such as may occur when an individual bends over suddenly or lifts a heavy object using the wrong technique.

What are the symptoms?

Symptoms may come on gradually or suddenly. The type of symptoms and the area affected depend on the site of the prolapse and whether the disc is pressing on nerves arising from the spinal cord or on the cord itself.

In lesions of the cervical spine where there is pressure on nerves, pain, tingling, numbness, and weakness may be present and particularly tend to affect the forearm and the hand. Movements of the spinal column are also restricted.

Lumbar disc lesions cause similar symptoms, but they affect the legs instead of the arms. The muscles of the back around the affected area may also spasm, as if to protect the vertebral column from further damage.

Occasionally, a lumbar disc ruptures and leaks its contents so that the cauda equina, a bunch of nerves originating in the lower part of the spine, is damaged. This causes a loss of bladder and bowel control. There may also be pain in the back that spreads down both legs.

How is it diagnosed?

A description of the symptoms and an examination will usually indicate whether a disc has prolapsed. An X ray will be done to look for other possible causes of back pain. An MRI or CT scan may be used to establish the exact site and extent of a prolapse.

What are the treatment options?

Cervical disc prolapses may improve with rest, painkillers, and sometimes traction. Wearing a support collar may also be helpful. In a few cases, the pain is so severe that surgery to remove the damaged disc may be considered. The relevant vertebrae may be fused to avoid any instability of the cervical vertebrae or persistence of the symptoms, although this restricts the movement of the cervical spine.

The symptoms of lumbar disc prolapses also often settle following a period of rest. Painkillers may be needed, and traction to stretch the spinal column and so reduce the pressure around the prolapsed disc may help, as may physical therapy. Back exercises will be recommended and should be continued for life to reduce the risk of recurrence. Anesthetic injections into the affected area may be given for persistent severe pain.

Surgery may be considered if the symptoms persist or are worsening, or if the pain is very severe. In the case of cauda equina lesions, urgent surgery is necessary.

Surgery involves removal of the disc, either by traditional surgery or by microdiscectomy, a newer keyhole procedure in which the disc is removed through a small incision, as the surgeon is being guided by images from a microscope introduced through another small incision. The recovery period tends to be shorter following microdiscectomy. Removal of a disc may put new stresses on the back, however, which may eventually cause stiffness and pain.

What is the outlook?

Although a disc prolapse may resolve with rest and painkillers, the condition has a tendency to recur. Even following surgery, symptoms may persist in some cases.

PSEUDOGOUT

Arthritis of a joint caused by the deposition of calcium pyrophosphate crystals.

This disease is similar to gout (p. 141) in the symptoms it produces. It tends to affect mainly women over age 60. Often there is a family history of the disease.

What are the causes?

The reason why pseudogout develops is unknown, but it sometimes follows an injury or an infection.

What are the symptoms?

The disease affects one joint at a time. The main symptoms are pain, which can be severe; inflammation; stiffness; and redness around the affected joint. Symptoms tend to occur in recurrent bouts.

How is it diagnosed?

Fluid may be withdrawn from the joint and examined under the microscope for the presence of crystals or tested for infection. Alternatively, the disease may be diagnosed from X rays that show particular changes.

What are the treatment options?

Nonsteroidal antiinflammatory drugs may be used. In some cases, pain may be relieved by removing fluid from the joint or a corticosteroid injection given into the joint.

What is the outlook?

There is no cure for pseudogout, although treatment usually relieves the symptoms. The condition tends to recur.

REPETITIVE STRAIN INJURY (RSI)

Muscle and tendon injuries caused by repeated use of a particular part of the body.

This is a relatively common problem; the muscles and tendons of the arms are often affected. Certain occupations, such as athlete, typist, and musician, can be associated with the condition.

Is RSI an inevitable part of keyboard work?

RSI is preventable and should not be considered an inevitable part of office or computer-based work. Experts recommend

- upright posture;
- not sitting in the same position for several hours;
- keeping the arms parallel to the desk or having the keyboard slightly below arm level;
- having the screen at or slightly below eye level;
- striking the keys with as little force as possible rather than "hammering" down on them;
- resting the hands and wrists whenever possible;
- resting the hand lightly on the mouse or trackball; and
- taking frequent short breaks: This is better than fewer long ones.

Ergonomic aids such as wrist rests, innovative mouse designs, and split keyboards may help, but only if an individual's posture and typing technique are good in the first place.

ASK THE EXPERT

What are the symptoms?

Tingling and aching may be accompanied by restricted movements in the affected part of the body. These symptoms come on gradually, at first occurring only during the activity but later being present at rest.

How is it diagnosed?

A description of the symptoms and an examination may be followed by blood tests and X rays to exclude other causes of pain, such as osteoarthritis.

What are the treatment options?

Taking breaks during the activity responsible for the condition is often advised, perhaps in combination with painkillers or nonsteroidal antiinflammatory drugs. Physical therapy may also be recommended. Changes may need to be made to the working environment to reduce strain; for example, adjusting the height of the desk and the position of the keyboard may help.

What is the outlook?

If the condition is diagnosed early and appropriate changes made, the prognosis is usually good.

RHEUMATOID ARTHRITIS

Chronic arthritis, usually affecting several joints in a symmetrical pattern.

In rheumatoid arthritis, the synovium, the membrane that lines joints, becomes inflamed. Eventually, the bone ends within the joint and the cartilage that covers them are also damaged. The disease can develop at any age and tends to affect premenopausal women more than men of a similar age. After menopause, women are affected at a similar rate as men.

What are the causes?

Rheumatoid arthritis is an autoimmune disease—the body forms antibodies that attack its own tissues, in this case the synovium. The reason why this occurs is not known.

What are the symptoms?

In most cases, this form of arthritis affects several joints, the symptoms developing gradually over weeks or months. Sometimes symptoms develop rapidly over a few days. Typically, the small joints of the hands and feet are affected first, followed by other joints in the body, which may include the wrists, elbows, shoulders, knees, and ankles. The hips are also affected, but less often. In some cases, just one joint is affected, in particular the shoulder or knee.

Joint symptoms include pain, swelling, and stiffness. The pain and stiffness are worse in the morning and may improve as the day goes on. Affected joints may also feel warm to the touch. Eventually, joint deformities develop as the damage within the joint worsens. The muscles around joints waste (become thinner) and lose strength.

The term rheumatism refers to any painful disorder of the joints or muscles not caused by injury or infection and so encompasses rheumatoid arthritis, osteoarthritis, rheumatic fever, and gout.

Other, less specific symptoms are often present, including fatigue (a result of the anemia that occurs) and weight loss. In addition, nodules may develop under the skin, particularly in areas that are under pressure. The elbows are common sites for these so-called subcutaneous nodules.

Rheumatoid arthritis may be associated with features in other parts of the body. Various disorders may develop in the lungs, including fibrosing alveolitis (a chronic condition in which the air sacs in the lungs, the alveoli, become inflamed and thickened, causing shortness of breath). The pericardium, the membrane that surrounds the heart, may become inflamed, a condition known as pericarditis. The walls of small blood vessels may become inflamed in a condition called vasculitis. The blood supply to the bowel or to areas of skin may be impaired. In addition, Raynaud's phenomenon, in which the arteries of the hands narrow when the hands are exposed to cold, causing tingling and numbness in the fingers, may develop. Less commonly, the spleen and lymph nodes in the area of the affected joint may become enlarged.

How is it diagnosed?

Blood tests to check for anemia and to measure the degree of inflammation (the ESR and CRP, p. 109) will be performed. The blood levels of rheumatoid factor, an antibody often found in people with rheumatoid arthritis, will also be measured. X rays may be arranged to look for evidence of the disease.

What are the treatment options?

Nonsteroidal antiinflammatory drugs are often prescribed to reduce the inflammation. In more severe cases, other drugs that reduce inflammation and limit joint damage may be considered. All of these drugs do have side effects, however, so it is a case of balancing the severity of the disease with the possible side effects. Corticosteroid injections given directly into the joint may be prescribed for severe joint pain.

Physical therapy is an important part of the treatment, as is support from doctors and specialist nurses. Surgery may be considered for severe disease. Possible procedures include an operation to remove very thickened synovium and complete joint replacement.

What is the outlook?

The pattern of the disease varies. In its mildest form, attacks affect one joint and last only a few days. However, this form tends to be recurrent and may progress to another form of the disease. In its most common form, the disease is a long-term problem, lasting many years, in which the symptoms occur in bouts with periods of remission. In some cases of rheumatoid arthritis, the symptoms last for about a year and then resolve without leaving any permanent damage. In others, the disease may last for a few years before it settles, leaving slight permanent damage. For a few individuals, rheumatoid arthritis is a rapidly progressive disease that causes deformity and disability in just a few years. This form is particularly likely to be associated with problems affecting other systems of the body.

About 25 percent of people make a complete recovery with no long-term problems; about 10 percent eventually develop severe disabilities that limit their activities.

RUPTURED TENDON

A tear in one of the bands connecting muscle to bone.

This is a relatively common condition. The Achilles tendon that extends from the calf to the back of the heel is particularly liable to rupture.

What are the causes?

The condition particularly affects athletic men and women. Sudden contraction of a muscle can cause a tendon to rupture—this is the most common cause. Less commonly, rupture results from damage caused by a deep cut or a fracture.

What are the symptoms?

The tendon is often felt snapping as it ruptures. Other symptoms include pain and swelling in the affected area and restriction of movement.

How is it diagnosed?

Investigations are rarely necessary; the diagnosis is made from a description of the symptoms and by examination.

What are the treatment options?

Nonsteroidal antiinflammatory drugs may be prescribed. In some cases, surgery may be needed to rejoin the two pieces of tendon. An Achilles tendon rupture may be treated by surgery followed by application of a cast to keep the tendon in position. Alternatively, a cast may be applied and the torn ends of the tendon allowed to heal without surgery. The cast holds the foot in a position that places minimal strain on the healing tendon. Physical therapy may be recommended for all tendon ruptures to help mobilization and to build up muscle strength.

What is the outlook?

Ruptured tendons usually heal well, but it may take several months to return to normal levels of exercise.

SCOLIOSIS

Sideways curvature of the spinal column.

Scoliosis can be seen when the spinal column is viewed from the back. It particularly affects females. The thoracic spine is most often involved, but abnormalities are also seen in the cervical and lumbar spine.

What are the causes?

Scoliosis may be caused by poor posture, sometimes as a result of a problem elsewhere in the body, such as one leg being shorter than the other. The cause of scoliosis is often unknown. The condition sometimes runs in families.

What are the symptoms?

The curvature may be present from birth or may come on gradually in childhood or later life. It often develops during adolescence at the time of the growth spurt. Back pain is common, and walking may be affected. In a few cases, the ribs are pulled out of position, and lung or heart problems may develop. Osteoarthritis (see p. 145) of the spinal column may eventually develop as a result of strain put on the joints by the abnormal position of the vertebrae.

How is it diagnosed?

The condition is usually obvious from an examination of the spinal column. This may be backed up with X rays.

What are the treatment options?
The underlying cause will be treated if possible. If the curvature is slight, no treatment may be needed and the patient will simply be reviewed regularly to see if the shape of the vertebral column changes. If the curvature is more severe, a brace may be used to try to keep it from worsening. For a severe curvature or for one that is worsening rapidly, surgery may be considered to straighten it or to prevent it from progressing further. The vertebrae may be held in place with metal rods or fused in the correct position.

What is the outlook?
Scoliosis may be mild and may not progress. In severe cases, however, it may cause postural, lung, and heart problems, and affected individuals may be distressed by their appearance. In such cases, surgery may be considered, but this will limit the movement of the vertebral column.

SEPTIC ARTHRITIS
Inflammation of a joint resulting from bacterial infection. The condition mainly affects children and the elderly.

What are the causes?
The disease is also known as infective arthritis. Infection may spread from a bone to the joint, resulting in septic arthritis. In some cases, infection travels to a joint via the blood, or it may enter if a wound penetrates a joint. People with rheumatoid arthritis (p. 150) are at an increased risk of septic arthritis, as are intravenous drug abusers.

Septic arthritis is relatively rare but is the most aggressive form of arthritis in destroying a joint.

What are the symptoms?
An infected joint is swollen, appears reddened, and feels warm. The condition causes severe pain. A person with septic arthritis is likely to be unwell and may have a fever.

How is it diagnosed?
A sample of fluid may be withdrawn from the joint using a needle and syringe and then is tested for infection.

What are the treatment options?
The joint cavity must be washed out, usually by running fluid into the joint and draining it out regularly. This must be combined with antibiotics, given intravenously for a few weeks and then continued orally to make sure the infection is completely cleared. Painkillers may be needed to relieve any discomfort, and rest initially is advised. Later, gentle exercise maintains the joint's movement and strength.

What is the outlook?
Without effective treatment, damage to the cartilage lining the joint will be severe and the joint movements will be greatly limited. When treatment is initiated promptly, recovery rates are good.

SPONDYLOLISTHESIS
Slipping forward of one of the vertebrae of the spine.

What are the causes?
Spondylolisthesis may develop for a number of reasons. An injury to the spinal column may cause one or more vertebrae to slip out of position. Vertebrae of the lower part of the back are usually affected. Vertebral fractures as a result of sudden stretching of the back, perhaps when pitching in baseball or throwing the javelin, are common causes in young people. Diseases of the bones such as osteoporosis (p. 146) and Paget's disease (p. 147) are possible causes, as is osteoarthritis (p. 145) in both men and women. There may be a congenital abnormality of the spinal column (one that is present from birth) where the lumbar and sacral parts meet. This rare condition affects girls more than boys.

What are the symptoms?
Often there are no symptoms. Pain and restriction of movement in the affected part of the spinal column may develop. Rarely, pressure on nerves branching off the spine causes pain to shoot down the leg from the lower back.

How is it diagnosed?
X rays of the spinal column are used to make the diagnosis. In addition a CT scan or an MRI may be performed to exclude other possible causes of the symptoms.

What are the treatment options?
A back support and reduction of activity levels are important, as are physical therapy and exercises to build up the strength of the muscles that support the spine. Painkillers may be needed. It may take several months for the condition to settle. If these conservative measures fail to solve the problem, an operation to fuse the affected vertebra in place may be considered.

SPORTS INJURIES

Damage to muscles, tendons, or bones that occurs while playing sports.

A wide variety of injuries can occur as a result of sporting activities. Muscle strains (p. 144) are very common. Muscle ruptures, in which muscle fibers are severed, are also relatively common, as is tendinitis (p. 154). Other possible injuries include fractures (p. 139) and injuries to the cruciate ligaments or cartilage of the knee (see Ligament injuries, p. 143, and Meniscal tears, p. 144). Fractures may occur as a result of a direct injury or from repeated stress on a bone. Joint dislocation (p. 138) is also common. Older people are at greater risk as flexibility and suppleness diminish.

What are the causes?

Sports injuries, such as a groin strain when playing football, may occur suddenly or may develop gradually. Repetitive strain injuries (p. 149) may develop when particular muscle groups are used repeatedly and sometimes excessively or with poor technique. Contact sports increase the risk of sudden injuries; rugby and football tackles are common causes. Some injures are specific to certain sports activities, such as the painful condition jogger's knee, which is caused by running long distances on hard ground.

Sports injuries often develop as a result of a lack of fitness or preparation. They often affect people who return to vigorous exercise after a break without pacing themselves adequately. Individuals who do not warm up sufficiently at the beginning of an exercise session may also be at risk.

What are the symptoms?

Muscle strains cause pain and sometimes stiffness. Ligaments, the tough bands that help keep bones in place, may also be damaged, causing pain. If a muscle or the point where a tendon meets a muscle (the musculotendinous junction) ruptures, a tearing sensation is felt in the affected area, often the calf or the thigh. This is followed by tenderness and later bruising, and, in the case of muscle rupture, swelling. Occasionally, a clot forms in the muscle, which prevents it from functioning properly. The risk of this happening is thought to be increased by starting to exercise too soon following a muscle rupture. The clot may gradually turn into bone tissue (ossify).

Tendons can rupture; the Achilles tendon—attached to the back of the ankle—is particularly at risk of this injury. A sharp, painful blow or snapping sensation is felt in the area of the tendon; this is followed by swelling.

Tendinitis is a common and painful condition; the patellar tendon of the knee is often affected in athletes who jump repeatedly, such as basketball players.

How are they diagnosed?

Many sports injuries can be diagnosed from a description of the symptoms and an examination. X rays may be needed to exclude fractures. In some cases, a CT scan or an MRI may be used to look at bones and surrounding soft tissues in more detail.

What are the treatment options?

Muscle strains and ligament damage may be treated by rest and sometimes by nonsteroidal antiinflammatory drugs. Returning to exercise should be a gradual process and may be helped by physical therapy.

Following a muscle rupture, ice wrapped in a towel should be pressed against the affected area to reduce swelling. The leg should be raised and rested for about six weeks. Ultrasound treatment may help reduce swelling during this healing period. The affected area should be moved manually to keep nearby joints mobile without

Can sports injuries be avoided?

Sometimes injuries are unavoidable. However, there are measures you can take to reduce the risks.

- **WARM UP** This should be a part of your routine every time you exercise and should include a range of muscle stretches. It is particularly important following a period of resting from your sport, when muscle strength and suppleness will have been reduced. At the end of your exercise, you should again follow a regimen of exercises to cool down and gently stretch the muscles. This will reduce the likelihood of stiffness and pain afterward.
- **WEAR THE RIGHT GEAR** Cushioned sports shoes and comfortable clothes that do not restrict movement are an important part of exercising. Adequate protection, including helmets, shin pads, and other equipment as appropriate, is also important.
- **EXPERT ADVICE** Seek advice on exercise programs that build up strength and stamina gradually and incorporate a warm-up session.

ASK THE EXPERT

contracting the affected muscle. Later, physical therapy and gradually increasing levels of use of the affected muscle are usually recommended.

Like a muscle rupture, a musculotendinous rupture requires a cold compress, elevation, and compression of the affected area (p. 96). Ultrasound treatment and manual movement of the limb will help the healing process.

For an Achilles tendon rupture to heal, the two ends of the tendon must be held together. This can be achieved by holding them in position with a cast or by surgery followed by the application of a cast. Ultrasound treatment accompanied by physical therapy may be helpful to reduce swelling and restore movement, but a gradual return to exercise is not recommended for at least 16 weeks.

Tendinitis is treated by rest and nonsteroidal antiinflammatory drugs. Jogger's knee may be improved by wearing running shoes with cushioned soles and by avoiding running on roads and other hard surfaces.

What is the outlook?

Muscle ruptures take at least six weeks to heal. Exercise should be avoided during that time and then may be gradually reintroduced with specialist advice. Even if care is taken, there is a risk of the rupture occurring again. A muscle clot that ossifies may disappear over a few years but in some cases may need to be surgically removed.

The pain following a tendon rupture usually lasts for about eight weeks. In most cases, the rupture heals well and normal activities can be gradually resumed. Achilles tendon rupture takes many weeks to heal and may recur.

Tendinitis settles over a period of a few weeks but can recur. Taking expert advice on sporting techniques may help to avoid the problem. Repetitive stress injuries can often be avoided or reduced by improving technique.

TENDINITIS

Inflammation of a tendon, one of the tough bands that connect muscles to bones.

The tendon sheath surrounding the tendon is often also inflamed, a condition known as tenosynovitis. A common site for tendinitis is the tendon attached to the patella (the kneecap) at the front of the knee. Other vulnerable areas include the shoulder, elbow, wrist, thigh, knee, and ankle.

Tendinitis is usually the result of placing repeated strain on a tendon, often during regular and sometimes excessive exercise. Tendon injuries may also cause tendinitis.

The major symptom is pain, which may be accompanied by swelling and stiffness in the affected area. Symptoms are usually relieved by rest and complete avoidance of the activity that sparked the problem, together with nonsteroidal antiinflammatory drugs. In a few cases, steroid injections are given around the tendon.

A program of physical therapy exercises will help to rebuild strength in the muscles surrounding the tendon, once the tendinitis has healed.

TENNIS ELBOW AND GOLFER'S ELBOW

Painful conditions caused by damage to a tendon near its site of attachment on the elbow.

What is the cause?

Tennis elbow is caused by damage to one of the tendons of the forearm where it attaches to the outer surface of the elbow. A tear may result from a sudden bending of the wrist. This movement may occur when hitting a tennis ball awkwardly, but it may also occur in other activities such as lifting or gardening.

Golfer's elbow differs in that it is the tendon attached to the inner border of the elbow that is damaged. This tends to occur when a golfer misses the ball and instead hits the ground hard. However, both conditions can come on gradually as a result of repetitive strain put on the tendons during exercise or other activities.

What are the symptoms?

An individual with tennis elbow has pain and tenderness on the outer border of the elbow. In golfer's elbow, the pain and tenderness is felt on the inside of the elbow.

How is it diagnosed?

Both conditions can be diagnosed from a description of the symptoms and an examination of the elbow joint.

What are the treatment options?

Both tennis elbow and golfer's elbow are treated by rest. Physical therapy may also be helpful, as may nonsteroidal antiinflammatory drugs. If the pain is persistent, a corticosteroid injection may be given into the tender area.

What is the outlook?

In most cases, the conditions settle. However, the problems may be recurrent and advice should be sought on how to avoid putting strain on the affected tendon.

TORTICOLLIS

Twisting of the neck as a result of muscle spasm.

Muscle spasm causes the neck to be twisted out of its usual position. The spasm is usually the result of sleeping in an awkward position or of an injury, often occurring during a sports activity. The condition can affect anyone.

In addition to its abnormal position, the neck is painful, and movement is severely restricted. The diagnosis is usually clear from the examination. Physical therapy may be recommended, although the symptoms often settle after a few days with rest and painkillers. A soft collar may be worn for a few days to provide support. A heating pad may also afford relief.

TRIGGER FINGER

Impaired ability to straighten a finger as a result of thickening of the tendon where it enters its sheath.

This is a common condition affecting one or more fingers. The thumb may also be affected.

What are the causes?

Trigger finger develops as a result of friction caused by the tendon at the point where it enters its sheath. A swelling develops on the tendon and prevents its normal smooth passage. The swelling can cause additional friction, and as a result, the entry point into the sheath narrows, making the problem even worse. Trigger finger is usually brought on by repetitive exercise.

What are the symptoms?

The muscles that bend the fingers are stronger than those that straighten them. This means that the affected fingers can bend but the straightening muscles are unable to pull the tendon back through the sheath. Thus the finger bends but then the swelling on the tendon catches and the finger cannot be straightened until the tendon is suddenly released and the finger snaps straight, or "triggers." Affected individuals often use their other hand to straighten the fingers.

The problem is usually worse in the morning after sleeping with the fingers bent and tends to improve over the course of the day.

How is it diagnosed?

The condition is usually obvious from a description of the symptoms and from the examination.

What are the treatment options?

Rest is required and will resolve the problem in most cases. For persistent trigger finger, a corticosteroid injection into the tendon sheath may be recommended. Occasionally, the opening into the tendon sheath is widened surgically to allow the tendon to pass through it more easily.

What is the outlook?

Most cases of trigger finger subside with rest and sometimes injections, with no need to resort to surgery. If, however, the operation is needed, it is usually effective.

WHIPLASH

Damage to the muscles and ligaments of the neck that occurs when the head is thrown back following a sudden impact.

What are the causes?

Whiplash is caused by the head being thrown first back and then forward, resulting in damage to the ligaments, muscles, and soft tissues of the neck. Being in a car that is hit from behind is the most common cause.

What are the symptoms?

Pain and stiffness of the neck are usually the main symptoms, although it may take at least 12 hours following the accident for them to develop.
An affected individual may experience discomfort when swallowing in addition to aching across the shoulders and down the arms. There may also be tingling in the arms, but this is usually short-lived.

Victims of whiplash lose an average of 8 weeks of work and are more often women. Whiplash can appear even after a collision at only 5 miles per hour.

What are the treatment options?

Wearing a neck brace for a few days is often recommended. It should not be worn for longer than two weeks, because the neck muscles will begin to weaken and stiffen. Pain relief is important. Following the initial healing period, physical therapy will help restore neck movement.

What is the outlook?

The prognosis of whiplash varies by individual. Most people find that the symptoms subside and that they have no persistent problems, although a small number—one study estimated less than 3 percent—are left with chronic pain.

Index

Acknowledgments

Carroll & Brown Limited would also like to thank:

Picture researcher
Sandra Schneider

Production manager
Karol Davies

Production controller
Nigel Reed

Computer management
Paul Stradling

Indexer
Jill Dormon

3-D anatomy
Mirashade/Matt Gould

Illustrators
Andy Baker, Rajeev Doshi/Regraphica, Kevin Jones Associates, Mikki Rain, Nick Veasey, John Woodcock

Layout and illustration assistance
Joanna Cameron

Photographers
Jules Selmes, David Yems

Photographic sources
SPL = Science Photo Library

7 Zephyr/SPL
8 *(top left)* Chris Bjornberg/SPL
(background) SPL
(bottom) Eye of Science/SPL
9 *(bottom right)* John Reader/SPL
10 *(top)* P Saada/Eurelios/SPL
(bottom) Michael Keller/Corbis
10–11 *(background)* SPL
11 Powerstock
12 *(top)* John Cole/SPL
(background) SPL
(bottom) Chris Bjornberg/SPL
13 James King-Holmes/SPL
16–17 Paul Steel/Corbis
21 Getty Images
25 Imagingbody.com
27 SPL
33 *(left)* Manfred Kage/SPL
(center) Astrid and Hans-Frieder Michler/SPL
(bottom) Innerspace Imaging/SPL
36–37 Getty Images
37 *(bottom)* Michael Keller/Corbis
39 Larry Williams/Corbis
40 *(left)* Philippe Petit-Mars/Corbis
(top right) Jules Selmes
(bottom right) Getty Images
42 Jules Selmes
44 Philippe Petit-Mars/Corbis
46, 51 Getty Images
53 *(top right)* Tom Stewart/Corbis
(4th from top) Wartenberg/Picture Press/Corbis
(6th from top) Getty Images
56 *(top)* Tom Stewart/Corbis
(bottom) John Henley/Corbis
64 Wartenberg/Picture Press/Corbis
69 Getty Images
71 *(left)*, 72 Carroll and Brown
73 Getty Images
75 Carroll and Brown
78 *(center below)* Getty Images
(right) Jennie Woodcock/ Reflections Photolibrary/Corbis
(bottom) Tom Stewart/Corbis
82, 83 Jules Selmes
85 *(top right)* Getty Images
(bottom right) Larry Williams/ Corbis
87 *(center)* Getty Images
(bottom) Rob Lewine/Corbis
93 Getty Images
94 *(bottom left, center)* Getty Images
97 Larry Williams/Corbis
98 *(left)* Dept. of Clinical Radiology, Salisbury District Hospital/SPL.
(center) BSIP Laurent H. Americain/SPL
(right) GJLP/SPL
99 Will and Deni Mcintyre/SPL
102 Dept. of Clinical Radiology, Salisbury District Hospital/SPL
105 *(top)*Mike Devlin/SPL
(bottom) Mauro Fermariello/SPL
106 Zephyr/SPL
107 SPL
108 *(top)* Catherine Ursillo/SPL
(bottom) SPL
109 J. C. Revy/SPL
111 *(left)* Carlos A. Wigderowitz
112 BSIP, TH Foto/SPL
113 *(top)* Getty Images
(bottom left, right) Terry Beddis/ Ricability
(bottom) Homecraftabilityone
115 Carlos A. Wigderowitz
116 *(top)* Jim Stevenson/SPL
(center, bottom) Dr. P. Marazzi/SPL
117 *(left)* Bill Longcore/SPL
(right) SPL
118 Damien Lovegrove/SPL
119 *(left)* Dr. P. Marazzi/SPL
(right) CNRI/SPL
120 *(left)* CNRI/SPL
(right) Zephyr/SPL
122–123 Carlos A. Wigderowitz
(bottom right) GCa/SPL
124 GJLP/SPL
125 *(left)* Oullette and Theroux, Publiphoto Diffusion/SPL
(right) Princess Margaret Rose Orthopaedic Hospital/SPL
126 BSIP Laurent H. Americain/SPL
128 *(top)* Simon Fraser, Hexham General Hospital/SPL
(center) St Bartholomew's Hospital, London/SPL
(bottom) Martin Dohrn/SPL
130 Will and Deni Mcintyre/SPL
131 BSIP Collet/SPL
132 Hattie Young/SPL

Back cover *(right)* Hattie Young/SPL

Contact details

American Academy of Orthopaedic Surgeons
1-847-823-7186
www.aaos.org

American Chiropractic Association
1-800-986-4636
www.amerchiro.org

American Orthopaedic Association
1-847-318-7330
www.aoassn.org

American Osteopathic Association
www.aoa-net.org

Arthritis Foundation
1-800-283-7800
www.arthritis.org

National Institute of Arthritis and Musculoskeletal and Skin Diseases
1-877-22-NIAMS
www.niams.nih.gov

National Osteoporosis Foundation
1-202-223-2226
www.nof.org